OMAR HATAMLEH • GEORGE TILESCH

BETWEENBRAINS

Omar Hatamleh & George A Tilesch: BetweenBrains

ISBN 978-1-7349318-2-2

9 781734 931822 >

The authors dedicate this work to humanity:
the once and future hero of Earth's story.
At this turning point in history, may the conscientiousness
of our better nature and the consciousness of
our best interests prevail.

THE AUTHORS

DR. OMAR HATAMLEH is the Former Chief Innovation Officer, Engineering at NASA, and former Executive Director of the Space Studies Program at International Space University. Prior to that he was the Deputy Chief Scientist at NASA ARC responsible for identifying new and promising areas of scientific research and supporting technologies that can be integrated into the Center's capabilities. Omar is also the President of the Infinity Institute, a global non-profit Think Tank tasked with developing disruptive solutions for healthcare. Omar has been at NASA for 23 years in different roles, and has published over 33 international journal articles in Engineering. Omar has been the recipient of several prestigious awards and recognitions from NASA including the Silver Achievement Medal, NASA Innovation, and Superior Achievement Awards. He has four engineering degrees, speaks four languages, and has been an invited keynote speaker to multiple national and international events including major innovation conferences, Google, IBM, European Investment Fund, etc.

DR. GEORGE A. TILESCH is a senior global innovation and AI expert who is a conduit and trusted advisor between the US and EU ecosystems, specializing in AI Ethics, Impact, Policy, and Governance. George has an extensive cross-sector and omni-stakeholder innovation background. His global senior executive and strategy consulting leadership track record span decades with the EU and US administrations; Microsoft and Fortune 50 Tech corporations; international organizations and global think tanks; pioneering startups/scaleups; and global social innovation leaders. Most recently, he was Chief Strategy & Innovation Officer for Global Affairs, the public interest arm of Ipsos, a global Top 3 research firm, where he led the Digital Impact & Governance research and advisory practice and coordinated AI thought leadership.

A senior consultant and frequent keynote speaker at major innovation conferences and a guest lecturer at US and EU universities, George is leading an AI Policy Working Group for Club De Madrid-World Leadership Alliance on AI, Trust & Democracy and is involved in multiple World Economic Forum Experts Groups, designing AI government strategy, and AI regulatory frameworks.

George earned a Master of International Relations and Advanced EU Studies degree from the European Institute, University of Nice-Sophia Antipolis, Nice, France, graduating with honors, and a Doctorate in Law and Political Studies degree from Janus Pannonius/Pecs University in Hungary. He lives with his family in Marin County, California.

TABLE OF CONTENTS

PROLOGUE TO THE FIRST EDITION

This book, written by friends, is an attempt to cover a topic as vast, complex and obscure as Artificial Intelligence (AI). We set out to accomplish that in a way that makes the domain approachable to responsible intellectuals worldwide: at the very least, we tried to ask the right questions and suggest a course of action. BetweenBrains was also inspired by our NASA cross-industry innovation community, that for many years have proved that the most in-depth, breakthrough innovation comes from other industries, sectors, disciplines and philosophies.

In line with this conviction, our work is going far beyond pure technology and is very much multi-disciplinary in its perspective. This is exactly the way we see AI emerging both as a cross-disciplinary field of study and as applied usage that cuts across artificial disciplinary borders and walks of life alike. Our mission is to bridge minds, state some hard truths and provoke thoughts for joint understanding and far better collaboration. We had to be mindful of necessary tradeoffs, so we ended up opting for cross-pollination and breadth to instigate a global conversation on AI. We believe that we can further deepen some of the most exciting directions in forthcoming volumes as deemed necessary by you, the kind reader. Also, our primary goal was to pinpoint all that is already happening now or is around the corner, so that we could build trajectories and scenarios based on the best data and wisdom that we currently possess.

Regarding the timeline being scrutinized, our view is that AI is here and is impacting our world to a great extent: therefore, we should focus on the now and the near-term. Unlike any previous waves of technology transformation, the depth of this coming one is unprecedented. AI is transformational across the board and makes us revisit fundamental questions of humanity. It will test our ability to adapt and may also trigger a higher level of attention to lessons from nature.

Beyond a pure technology lens, we believe that the AI aspects that deserve most of our attention are its power and impact. We are approaching this in a balanced way, both philosophically and through demonstrating applied AI use cases in a wide range of industries. Talking about our emphasis on balance, we are convinced that both optimism and pessimism are futile terms to associate with AI: we have to uncover the truth and for that, cut through a dense jungle of obfuscation and distrust. With great potential comes great risk and both need deep understanding, new structures of trust and hard work for all of us.

In its present incarnations, AI may be narrow and brittle, but already deployed widely. The public worldwide has reacted to it with curiosity, concern and confusion. It is known to be very alluring to those who understand its promise and can also foot the bill. We are convinced that AI will serve as the most important instrument of power redistribution in the 21st century. Moreover, AI is the core of exponential transformation: soon all technology will be AI-powered.

The primary question of our age is: who gets to shape AI? To be prepared and to be able to participate in steering its course, one should learn its nature, power principles, state of play, and plausible scenarios. In addition to a sweeping analysis, we explore the realms of policy, war and religion as future case studies where opportunities abound but understanding tradeoffs and clear warning signs is critical.

As the first "moral technology", AI is not neutral. The ethical consequences of AI development are playing catchup with those who dictate the fastest possible pace in both research and deployment. We will put the limelight on these ethical aspects and cutting-edge research directions that would ensure a humanistic course for machine intelligence. Most pragmatically, in the timeliest subdomain of AI impact on democratic institutions and the public dialogue, we even attempted to include policy recommendations that may have relevance in other subdomains as well.

This book would not be wholesome without a solution framework that gets humanity prepared for AI-readiness. We are submitting a comprehensive vision, principles and institutional building blocks for constructive, global debate and implementation: it is in the shape of a manifesto that concludes this book.

While we are no neuroscientists, our title was originally borrowed from neuroscience: the betweenbrain (diencephalon) is a small but important part of the central nervous system. It is a collection of structures that play a number of critical roles in our health, including functions as diverse as circadian rhythms (sleep and consciousness) regulation, reward processing, movement modulation, sensory data processing, hormone secretion and most importantly, maintaining homeostasis - system equilibrium in the body. On top of that, in our intention, the title hints at the intricate interplay between human and machine intelligence that, we believe, will define our future to a great extent. In this setup, contrary to the dominant narrative, nothing is inevitable. The "inbetween-ness" of our age also equals an opportunity to make clear choices on what AI objectives we as a society want to achieve, what should be its public-interest purpose, and what choices and actions will take us there. Ultimately, we are firm proponents of Augmented Intelligence as the motto – the less "artificial" it is, the better off human civilization will be.

We can only hope that the reader will feel more clear-sighted and better equipped for the AI Age once our work has been perused. We wish you an enjoyable reading time: please do not hold back your imagination.

Omar & George

March 2020

CHAPTER 1

ON ROBONOMICS AND THE FUTURE OF HUMANITY

As children, we are all asked, "What do you want to be when you grow up?" We think of different things at different times, depending on what excites us, what is fashionable, what might impress other people. Perhaps you went on to do exactly what you hoped you would – but did it happen in exactly the way you expected? Imagining the future is not easy, and in some ways in this book we are asking what humanity wants to be when it grows up. We see some crucial choices coming, and some decisions that have already been made on our behalf. We want individuals, families, companies, societies, and states to take an active role in what comes next. Why? Because the new wave of change enabled by Artificial Intelligence (AI) will have a profound impact on society, our jobs, our health, and our politics.

What are the challenges and opportunities AI brings? As we will see, some are being addressed at different levels – by companies wanting to make money, by researchers looking for better solutions for everyday problems, by countries and organizations hoping to maintain society's values in a parallel digital world. But there is more than just AI to take into account. For example, until now, no significant discussions have taken place regarding how population increases, and exponential technological advancement taken together have created a perfect storm for a shift and disruption to our future society and economic models. The narrative in the following chapters will touch on just a few of the seismic effects that AI could bring today and in the coming decades.

Do you remember your day-to-day life when you first started school or started working? How many of the tools you use every day now existed then? Now think what a child born in 2020 will see in their lifetime. It's hard to imagine what the world will look like in 20 or 50 years, given the unprecedented acceleration in technological development and adoption, especially in the fields of data science and computational power. In stark contrast, if we turn the clock back 2,000 years, a prediction of what the world would look like in 50, 100, or even 1,000 years from that time would have been a relatively simple task.

Take for example, preeminent Greek philosopher and scientist Aristotle, who lived 2,400 years ago. His philosophy exerted a unique influence on almost every form of knowledge and his introduction of the method of scientific thinking is still relevant today. He had an especially brilliant mind but was working in a time when change was slower. If we consider the state of technological maturity of those times, Aristotle may have been able to imagine – and even adapt to – the way of living in the world 1,300 years after his time, when another great polymath, Ibn Sina (Avicenna), lived. Ibn Sina was regarded as one of the most influential physicians, thinkers, and astronomers of his time, and considered by many to be the father of modern medicine. We think Ibn Sina would have been able to envision and adapt to life in the 15th century, had he suddenly found himself a few hundred years into the future.

Perhaps the visionary who was most ahead of his time in predicting and designing early technological concepts several centuries ahead of his era was Leonardo da Vinci. The Italian polymath was born in 1452 and has long been regarded as the model Renaissance man and a prolific inventor. Leonardo was revered for his technological ingenuity. He conceptualized a bewildering number of inventions including technologies like flying machines, and armored fighting vehicles. Da Vinci may have also designed what may have been the very first humanoid robot by mobilizing a mechanized armored knight using gears and wheels connected to a pulley that enabled the basic "robot" to perform limited movements. However, due to the lack of scientific technology and advanced materials, his inventions were not feasible during that era, and his technological concepts remained unmaterialized. We might not be as creative as these individuals, but we are going to need to start thinking as they did to adapt to the pace of change that is coming.

We have advanced so much as a modern civilization that it is difficult to make sense of the fact that, just 100 years ago, Western society used horses as the primary means of transportation. Labor-saving machines brought radical change to factories and homes in the 19th and early 20th century, and in recent decades computers have been having a similar influence.

Unprecedented from the previous 2,000 years of human technological evolution, the present times are characterized by rapid, non-linear improvements in enabling technologies such as AI and Machine Learning (ML). ML is a branch of artificial intelligence that is based on the premise that the system can learn from data.

Computing machines date back to the abacus and took a leap forward with Charles Babbage's Difference Engine and Analytical Engine. But it is really only in the last 90 years that modern computers were born and grew up with astonishing speed. Computer chips are increasing in power while the cost is diminishing. This is attributed to the fact that the number of transistors on a computer chip has been doubling roughly every 18 months. AI enthusiasts like Ray Kurzweil believe that the cognitive ability of a computer will soon be equivalent to that of a human. This will enable a computer to achieve complex tasks, such as performing medical diagnoses, creating music, or arguing with an opponent in a political debate. We need to prepare now for the moment when computers may become our equals, and maybe even outstrip our achievements.

What are the social and economic implications of such a paradigm shift in technological capabilities? In a few decades from now, if robots take on the form of humans, it may be hard to distinguish a humanoid robot from a human. Now imagine the possibilities and the societal implications of having advanced humanoid robots. What if they become as technically and cognitively capable as humans (or even superior to them) in several economically and socially relevant tasks and contexts? These technologies and emerging capabilities will significantly impact our society and the job markets of the future.

The establishment of shared theoretical frameworks, combined with the availability of data and processing power, has yielded remarkable successes in various component tasks such as speech recognition, image classification, autonomous vehicles, machine translation, and question-answering systems.

> "As capabilities in these areas and others cross the threshold from laboratory research to economically valuable technologies, a virtuous cycle takes hold whereby even small improvements in performance are worth large sums of money, prompting greater investments in research. There is now a broad consensus that AI research is progressing steadily, and that its impact on society is likely to increase [...]. Because of the great potential of AI, it is important to research how to reap its benefits while avoiding potential pitfalls."

From an Open Letter, Research Priorities for Robust and Beneficial Artificial Intelligence, January 2015[1]

There are discordant views about AI among scientists, techno-futurists and economists that are polarizing those in the general public. Some say that various forms of advanced robots and AI will create new and diverse job opportunities, while others argue that they will have devastating effects on the job markets of the future. Similar thoughts were circulated before the Industrial Revolution, when the majority believed that automation would devastate the job market; this turned out not to be the case.

The difference now is that we are not only competing for manual labor, but also for highly skilled jobs requiring intellectual skills and capability. Another important point to consider is that, for the first time in human history, value creation and human labor are separated.

Making Sense of AI

How will individuals and society as a whole evolve, and possibly (be) transform(ed) by these emerging and potentially pervasive technological capabilities? Should we expect co-evolution and cooperation between humans and AI-empowered humanoids or instead competition and potential conflicts?

Some of the answers to these profound, multi-faceted, and daunting questions will likely depend on if, when, and mostly how AI will reach levels of complexity, autonomous reasoning, and behavior(s) to de facto result in full AI capabilities of self-determination.

At this infant stage of AI evolution, humans are still playing a central and paradigmatic role in how they can conceive, design and shape the way of thinking of AI, from processing data to making sense of complex information and possibly acquiring real knowledge. In the future, autonomous robots driven by AI may even have a collective intellect and learning capabilities as they may be connected to each other. Whatever one robot learns, so do all others. That will substantially increase intellect and accelerate the rate of cognition in a way that is unparalleled to anything we have seen to date. What role will humans have in a not-too-distant future compared with intelligent, capable, and autonomous systems?

This is something of an existential challenge to humans and has sparked many concerns, hopes, and hypes surrounding AI. It's crucial for the general public to clearly understand the distinctions between different levels of sophistication of AI to neither underestimate nor overestimate its capabilities. This will enable individuals and society to better assess the potential threats and benefits brought forward by AI in the future.

AI originally referred to the attempt to emulate a human-like form of intelligence, in all or at least most of its nuances. The term "AI" is often used in reference to powerful tools for data classification. These tools are impressive, but for now, they are on a totally different spectrum than human cognition. One of the fundamental challenges in emulating human intelligence is related to the fact that despite our extraordinary advances we do not fully understand how the human brain works. Consequently, we are not aware of any software, complex set of algorithms or AI platform that is even remotely as intelligent as the most cognitively challenged human being. Strictly speaking, no one has built (yet) anything close to real human-like AI capabilities on Earth.

At the current stage, AI is categorized as Artificial Narrow Intelligence (ANI or Narrow AI), which is limited in scope and restricted to only one functional area. ANI has the ability to outperform humans at some narrowly defined job. Two more stages – as yet theoretical – have been defined. The second stage of AI is Artificial General Intelligence (AGI). AGI also known as human-level AI can understand and reason in its environment as humans do and covers several fields that are on a par with adults. The third stage is Artificial Super Intelligence (ASI) in which AI surpasses human intelligence across practically all fields. Later on, we will see how industry, policy and society may be affected by each of these stages, and consider what actions might be sensible, or indeed valuable, to take now.

The quest for mimicry of intelligence has inspired the imagination of the general public for several decades. AI has been discussed for decades so why has it suddenly become one of the pervasive technologies that everyone is talking about? That is attributed to three main pillars.

First Pillar – the evolution of the fundamentals of neural networks (NN) and their associated complex algorithms. Early models for NN were single-layered and very basic, unable to perform complex cognitive tasks. Recent advancements in the field have resulted in more complex multi-layered NN that are significantly superior to the early models. These advanced algorithms are better suited to transcribing speech, and identifying objects including handwritten digits. Also, these advances in NN made it easier for the algorithms to recognize objects by looking at them from different angles.

Second Pillar – the video gaming industry had a significant impact by developing superior hardware to perform faster calculations by running calculations serially, thus allowing for more computations to occur simultaneously. It is astounding that in the 1960s when the integrated circuit (IC) was invented, an average IC contained two transistors. Today's most sophisticated ICs comprise over 20 billion transistors.

Conventional Central Processing Units (CPUs) in PCs such as those manufactured by companies like Intel have been designed to run calculations serially. The rise of dual and quad-core CPUs has enabled the capability of the CPUs to perform more computations to occur simultaneously. However, current Graphics Processing Units (GPU) can contain several thousand cores in comparison, allowing them to process substantial amounts of data at the same time.

The CEO of NVIDIA predicts that GPUs will even get a thousand times faster by 2025 compared with the current ones. This will result in substantial benefits spanning across industries like healthcare, energy, cybersecurity, automotive, and financial services. The computational capabilities of parallel processors have advanced NN and machine learning and have substantially increased the processing speed for current existing AI algorithms, opening up various possibilities to expand their use across multiple platforms and technologies.

Third Pillar – availability of Big Data. AI requires data to build its capability, in particular, for machine learning. For example, a machine learning image recognition system, evaluates thousands of images of an object to understand and recognize later on. Therefore, Big Data derived from reliable and bias-free datasets can deliver the information required to train the algorithms to perform a superior function. AI algorithms continue to absorb new information and data to modify and adjust their results, mainly through trial and error. Therefore, the greater amount of data an AI application has access to, the more accurate the outcome it produces. In the past, AI did not work well due to unsophisticated and primitive NN algorithms, slow computational capabilities, and insufficient/scarce/limited data sets.

The combination of these three pillars has enabled the advancements of AI, creating a remarkable field. But AI is not the only technology to have enjoyed big strides forward.

Imagine the possibilities created by combining AI with other emerging exponential technologies like autonomous systems, drones, quantum computing, biotech etc. These intersections of fields are enabling and creating waves of new capabilities able to transform and disrupt industries across the spectrum. The possibilities of these waves of capabilities will be even more pronounced when used across various industries, generating new synergies and value. Domain-specific AI-based expert systems have already started to deliver tangible benefits and economic value in a wide range of traditional industries. However, in our opinion it is inaccurate to suggest that at the present time, machines (computers and robots) have emulated the intricacies of our cognitive processes. Some current applications around AI and humanoid robots, are emulating behaviors that resemble intelligence but are not real intelligence. These systems are perhaps effective in a specific domain, but very limited, and inadequate in others. While it may be too early to declare that AI has equaled human cognition, it is certainly not too early to prepare for that moment.

What impact will exponential technological advances have on the future of jobs and future economies? There is huge potential for good, but there are also threats to jobs, especially as the world's population increases over the next decades.

Future impacts of technology on human-centered jobs could be amplified by the acceleration in population growth on a global scale. If you look at the Earth's population for the several hundred thousand years until the 1800s, the total number of people on the planet was less than one billion, and the average life span for humans was in the 40s. In the span of two hundred years, we went from less than a billion to about 7.5 billion people and the average life span is now in the mid-70s. That increase has been mainly attributed to clean water, vaccines, antibiotics, and other advances in medicine.

The UN estimates that the world population will reach close to 10 billion people by the year 2050. But is that large increase big enough? In our opinion, that estimate does not take into account the technological breakthroughs and advances in the fields of biotech, nanobots, and artificial intelligence that are expected to increase the average lifespan. Several studies are emerging related to anti-aging gene therapy to address aging in animals. Gene therapy could one day help stall the development of multiple age-related diseases and extend human life span. A recent study by Harvard University indicated that gene therapy was successful in dramatically reducing multiple age-related diseases in mice, and further studies are extending to other animals like dogs as well. The results suggest that these advanced therapies could improve overall health and lifetime.

Another recent study by Cardiff University discovered a method of killing several kinds of cancer in laboratory tests. The team discovered that a single type of an immune cell called a T-cell, and its receptor, could be adapted to find and kill a wide range of cancerous cells in labs and is showing enormous potential for curing all types of cancers in humans. Advances in 3D bioprinting are also making long strides and may be able to produce functioning whole organs in the not-so-distant future. These organs may even be based on the genetic compositions of the recipients by using individualized stem cells, preventing any possible rejection.

Advances in AI will also have a big impact on advancing the state of the art in pharmaceutical applications and helping us to understand the fundamentals of human aging. The promising results from gene therapy, biotech, and other technologies will mark a milestone to advance science and help treat many diseases. These advances may even lead to a resource to address aging itself, extending the average human life span well over current possibilities. This will translate into a world population beyond what is currently estimated.

What will the future economy look like with a large increase in population and potentially much less roles for humans in future job markets? How will that impact a retirement system that is already strained? Will working hours decrease as we move forward, to potentially only a few hours per week? Will a universal income model eventually be adopted, whereby robots will work, and humans will be free to do other things? All these questions must be explored in detail; we need to put our collective intellect together to start strategizing about this possible paradigm shift.

Our Social Fabric

Our society is already experiencing what we in this book will refer to as **Phase 1**. It is a phase resulting in diminished human-to-human personal interactions and communications, and instead shifting to communications using electronic media. This phase is particularly evident in younger generations who seem more comfortable communicating with digital services and devices like social media instead of face-to-face dialogues and interactions.

But we believe this will evolve into a second phase. In **Phase 2**, people will transition from spending more time with other humans over electronic media to devoting a significant amount of their time communicating with an AI-based system or humanoid robot within an advanced AI ecosystem. They may prefer machine friends to flesh and blood friends. AI algorithms can identify small changes in preference and adapt to users' needs by using data collected from users over extended periods. Soon AI will be so advanced that it will probably know more about users than they know about themselves! This will build a strong bond between AI and humans with a tremendous impact on social behavior and interactions.

But what if that system was hacked or optimized to have an influence on you? For example, to persuade you to buy certain products, change your political views or manipulate you to hand over more and more data, the same way as the current data-exploitative business models work? Left unchecked, abuses of technology in Phase 2 could have severe social, political, and economic consequences. The social implications of this phase of technology on society and relationships are still to be seen. Humans evolved to live in communities, develop social relationships, social interaction, and culture as a community. Taking that away would fundamentally change the essence of society and humanity as a whole. Would we even notice it happening?

Another aspect will be the effect of virtual reality integrated with AI. People will devote more time to using virtual reality powered with advanced AI in which a parallel world is created similar to Second Life. Second Life is an online virtual world where users can create content, interact with each other, and have a simulated virtual life. Advances in AI and Virtual Reality will take this technology to a very different level. As virtual environments and landscapes increase in popularity, a new economic model will develop, driven by the sales and acquisition of valuable virtual real estate. Another concept to discuss is Telexistence, first proposed in Japan in the 1980s. The concept describes the ability to have a human being experience a real-time sensation of being in and interacting with a remote location and environment. By combining virtual reality, AI, and robots, it is now possible to use these technologies to create an avatar-like concept that enables the user to have real-time interaction and immersion with a remote environment using a robot that resides remotely. Telexistence Inc. has recently started to industrialize the Telexistence concept, elevating the technology to a different level. This technology will keep maturing, eventually creating new applications in fields like education, medicine, and tourism. The more these virtual reality worlds are integrated into our everyday lives, the more important it will be to ensure that these environments are safe and secure.

Will this new relationship with technology result in humans becoming disconnected/disengaged from society? Perhaps. A technocratic society, dominated by pervasive AI taking away jobs from humans, is only one of the possible future scenarios that can unfold in the coming decades. We can model and convince ourselves of many other plausible futures. A potential outcome is a future of non-competitive abundant resources shared in a fair way by all people on planet Earth. On the other extreme of the quantum spectrum lies a future of artificial scarcity, poverty for most, injustice, and deprivation, in which fundamental freedoms and human rights are not guaranteed. There is a future in which people could thrive on innovation and prosperity, working just a few hours a week, with amplified opportunities and possibilities, like a modern Renaissance; or a future of jobless humans, subdued and oppressed by autocratic intelligent machines with no respect for human nature and values. All these futures are plausible and within reach, given our current technological and cognitive capabilities.

The long-term impact of AI on the economic models of the future should be looked at carefully, but what about the short-term implications? How can corporations, governments, and academic institutions adapt to a new world where AI will start imparting tectonic shifts by radically transforming the conventional operating models? How can they adapt and transform to remain relevant and effective? What will the emerging technology ecosystems be in the upcoming decade?

There is an abundance of synergies between nature, business models, technologies like AI, and future industry models. We are the intelligence that nature created, so how will nature react to the intelligence we create? Nature tends to favor equilibrium: we believe that technologies and societal evolution that go against the underlying rules of nature will not last and will inevitably collapse. To elucidate this connection between nature and technology, we want to discuss some lessons from nature, and how they can offer the key to design, interpret, and make sense of the technological societies of the future. And most importantly, how to remain relevant in an ever-changing landscape.

How Does Nature teach us to thrive in the Age of AI?

Adaptation

Business models need to be adapted and modified to fit the changing landscapes of the future. Adhering to the current models will not fit the new emerging landscapes and will result in failures. To bring an analog from nature, a prime example is polar bears that evolved from common brown bears. It is thought that polar bears' ancestors (brown bears) migrated to the Arctic due to changing global climate and increasingly warmer weather thousands of years ago. In the new environment polar bears had to quickly adapt to their new Arctic habitat. Through evolution, polar bears developed their white coats, which allowed them to blend seamlessly into the Arctic environment and enable them to be better hunters and collect food, as it was more difficult to be identified by other animals lower on the food chain. Adaptation allowed them to survive and thrive. The days when an engineer graduated from college with a degree and never went back to college are over. Successful employees of the future are the ones that are easy to retrain, adapt, continue to learn and engage in the ever-changing new ecosystem. AI can help organizations adapt and transform businesses to better fit in future landscapes.

Executives will need to figure out how to capitalize on AI to transform and bring forward progress that will enable corporations, academia, and government to generate creative processes and higher profitability. When a group of executives or decision makers from certain industries tries to set the strategic direction of the company or predict the future implications of technologies, they need to step back and analyze things from a distance. They should do so in a way that has a cross-industry component and they should analyze the societal impacts of emerging technologies like artificial intelligence and machine learning on our lives. In doing so, executives will implement an approach to the future impacts of technologies in a way that reduces uncertainty and factors in an awareness of the unknown.

AI systems are not just tools of adaptation for businesses. They also have the ability to learn and adapt as they generate value and make decisions. By taking advantage of AI's ability to utilize vast amount of data to learn, and identify emerging trends, users can leverage AI to position their industry for success in this new world. AI can help in drafting a future strategy, by scanning thousands of relevant parameters and anticipating probable scenarios, prompting strategists to adapt and transform their organizations to remain competitive and relevant.

Be part of the Future Ecosystem

Recent research has indicated that trees share resources and communicate with each other underground through a network of fungus called mycorrhizal fungi. The fungi and the trees share a symbiotic relationship and are interdependent on each other. This fungal network that connects the vast majority of the plants has been nicknamed the "Wood Wide Web" by scientists. Through that complex network, trees can share nutrients with others and "communicate" with them about possible dangers and plagues like insect infestations. Some plant species even leverage that system to support their offspring and incapacitate their possible adversaries. This understanding is challenging scientists to classify forests as a single superorganism as opposed to a combination of independent and individual organisms.

Government, industry, and academia can learn a valuable lesson from this, and work with each other in a cross-industry way. The boundaries between industries are becoming blurred with the emergence of new technologies like AI. AI will have applicability and value across different industries and will have implications for most business lines. The value will range from education, healthcare, finance, to manufacturing, and beyond. Where do you see yourself fitting in?

AI will even gain more power when combined with other emerging technologies like the Internet of Things (IoT), a concept describing physical objects being connected online with the capability to identify themselves to other devices. It is expected that in the year 2020 the IoT connected devices worldwide will number close to 21 billion, doubling in number a few years later. AI will extend the use of IoT from conventional applications like smart locks, smoke alarms, security systems, etc. to introducing more robust cyber security capabilities, and functions incorporated in future smart cities. The synergy and combination between AI and IoT will optimize the operational flow and productivity across industries, generate useful analytics, customize user experiences, and gain relevant insights by intelligently interacting across different ecosystems. Things will start to work together and may not rely on our input and willinstead, anticipate our needs.

There is room to exploit the value from the ecosystem across several sectors. For example, AI will be the center point in developing the capability for autonomous vehicles and trucks, and power the brains driving humanoid robots with applications that include healthcare, robotic surgeries, home automation, and several military applications.

Autonomous cars are being trained to drive using conventional driving scenarios. In a benefit from cross-industry knowledge transfer, vetted advanced algorithms developed for autonomous racing cars like the ones developed by Roborace can be incorporated into autonomous driving ambulances that can get to the destination in a shorter time.

Future technologies will also have applications across multiple industries. Establishing alliances and partnerships with industries across the board will produce large ecosystems with benefits to all. AI will have the potential to transform businesses and enhance economic growth at a higher scale.

Continuous Renewal of Operating Models

The human body is a complex machine that is constantly repairing and renewing its cells. Most of the cells in the human body go through constant renewal. Hair falls and new ones appear, skin cells are constantly shed and replaced with new ones, red blood cells are renewed through our bloodstream roughly every four months. This renewal, rejuvenation, and replacement rate of many of the human cell types is the norm. Future technologies will be ever changing and updating at a fast pace as technology stays on the path of exponential updates. Business models need to keep updating to reflect the reality on the ground.

Many examples exist of corporations that not only renewed business models and disrupted their markets, but also kept updating their business models to continue to be relevant. Examples include companies like Netflix, Airbnb, and Uber. Strategic direction of corporations should take this into account to develop AI strategy and an adaptive culture in which business models are adapted to upcoming technological trends, and to enable corporations to remain relevant to their customers and competitive.

Advanced algorithms can help draft and augment business models based on thousands of business cases and unique conditions that are relevant to the specific organization.

For example, AI can be used to radically improve marketing and customer service, and to optimize supply chain management. Several businesses, governmental institutions, and academic institutions are already using AI to improve operations and value. All new organizational business models will need to take AI into account and capitalize on the great opportunities it can bring. This will enable them to continuously change and adapt to remain competitive in the current wave of fast technological and societal transformation.

Exploring New Markets while Preserving your Core Business

Certain bird species are known to migrate and move from areas of diminishing resources to areas of mounting resources. These birds tend to migrate in the spring to benefit from the increase in insect populations and from the vast nesting locations in other regions. The same birds start migrating back as winter approaches and food resources become scarce.

Scientists believe that long-distance migrations may be coordinated partially by the bird's genetic composition. Other factors like weather, food sources, and day length are also believed to have a significant influence on migration patterns. The seasonal abundance of food and longer days are thought to enable birds to raise more chicks compared with other birds that did not migrate.

Corporations need to constantly explore other markets and revenue streams outside of their traditional ones – even across different industries and different geographical locations. One example is Amazon. It is interesting to follow the progression of Amazon from an online retailer to a leader in web hosting services, banking, education, entertainment, etc. By diversifying their portfolio and exploring new markets, they ensure a diversity that will solidify their long-term viability, while embracing future technology ecosystems.

To be the best and most successful, corporations need to anticipate what new technologies will have big impacts in the next 5–10 years and start investing resources in these technologies before they become mainstream. Doing so will ensure leadership in the market. Another example is the investment by Google, Microsoft, Intel, and IBM in early research and development of quantum computing, which could power vast advances in artificial intelligence.

Human intuition about the future is linear, so unfortunately, decision makers and strategists can fall victim to linear bias and linear thinking in a nonlinear technological world. Advanced AI algorithms can also help decision makers overcome these pitfalls by providing accurate trends and powerful insights. When industries strategize about future products and markets, strategists should not anchor their thoughts with the limitations of today's technologies. AI systems will be able to research what the trends are in patents in certain industry segments, what the latest research reports in that field are, what venture capitalists are investing in, the trends in number of graduates, how many startups are created, how the market is, the geopolitical status, and how public policies are changing. All of these parameters will allow strategists and technologists to better connect the dots and provide themselves with a better understanding of what technologies and products may be mainstream in the next decade. AI and Big Data can be of tremendous help in anticipating what future ecosystems will be relevant. This will allow companies to develop the capability to transform industries to become the market leader due to early adoption and capacity building.

The complex analytics generated by AI will empower more accurate predictions and yield better recommendations about future markets and trends. Technologies like adaptive machine learning can also help create sophisticated analytics that can be very helpful in assessing data or content to complement traditional business insights.

Exaptation

Exaptation is typically used to describe a change in the function of a feature during the evolution process. For example, a trait can evolve from serving a particular function to subsequently serve another.

Exaptation is shared in both anatomy and behavior. An example of anatomy exaptation includes the co-option of feathers in birds, initially evolved for the purpose of temperature regulation and display but later adapted to enable birds to fly. Trees may have evolved to produce sweet fruits to make them desirable and craved by animals and humans. By doing so humans and animals will disperse their seeds and ensure the survival of the plant's species.

Exaptation is not only applicable in biological evolution but also in technological innovation. In an era of technological abundance, it is important that organizations use innovation and exaptation to create other functions and models from existing ones used for other purposes. Microwave ovens for example were invented by chance when Percy Spencer noticed that a candy bar in his pocket melted while working on building a magnetron for radars sets. That technology was later exapted for food cooking and heating.

Currently there is an abundance of intellectual property that can be repurposed or used in areas and functions outside of their original intended application. By using these ideas, various industries across the board will benefit substantially from a fresh infusion of technologies that are exapted from their original intended use and can create vast economic value. Using advanced AI and machine-learning algorithms can help this process: these systems can identify other potential uses and value for existing technologies or products outside of their intended purpose or domain. This will generate enormous value if exploited properly.

AI will offer a wide range of tools helping to evaluate and produce better information integration, and data analysis to improve decision-making. This can be done by leveraging data collected from various sources, across different platforms and spectrums. This will allow AI to offer better capability of tremendous sophistication and analytics.

The key to better analysis of sophisticated trends and interconnections is through taking advantage of powerful AI capabilities. By having access to enormous amounts of data from different industries and ecosystems like finance, healthcare, smart cities, IoT, etc. AI will be able to find value in areas across various unrelated fields, and exapt or modify its capability to produce superior results in their field of application. Unsupervised learning algorithms, for example, can be used to help find previously unknown patterns in datasets, yielding great insight to users.

Failures can lead to new business evolutions and inventions

Investigations into a variety of insect species have indicated that DNA errors can have beneficial consequences. Monarch butterflies for example have shown that genetic errors have enabled them to consume highly toxic plants like milkweed in their larval stage and adapt the toxins to defend themselves against their natural predators.

There are several institutional lessons that can be learned from failing and making errors, but also in some cases, revolutionary inventions can be created. For example, Charles Goodyear did dozens of experiments to try to make rubber weatherproof. His victory came by combining rubber and sulfur and accidentally leaving it on the stove, although he later lost out on the material rewards of discovering this process, now known as vulcanization.

Failures and errors can provide valuable knowledge that can be used in a positive way. The knowledge that is gained from failure and errors can be quite important and may have a big impact on an organization. Failure can augment our experience, knowledge, and resilience. As long as we learn, adapt, and don't repeat the same errors, failure may be an option.

AI can play a major role in knowledge management and help transpose the failures into properly characterized lessons learned, which could bring great benefits. Current machine learning algorithms are proficient at training themselves and learning from their failures, thanks to the process of reinforcement learning. Reflect on how humans learn from failures: when playing a sport, playing an instrument, or learning how to drive. Initial failures enable us to adapt and adjust. With repeated failures, we get closer to our goal and achieve success. This method enables the AI to learn from its own mistakes, similar to how humans learn from their mistakes. By doing so, AI will continue to learn and add more to its capability, offering an incredible potential with the power to create waves of capabilities in many different fields.

Small changes can have a big impact long term

New research has shown concrete evidence that DNA can be transferred between different species. Specifically, the DNA transfer between some parasites and their vertebrate hosts.

For example, humans may be carrying DNA from certain parasitic insects, possibly resulting in alterations to human evolution over time.

The transferred genes were transposons, which are the segments of DNA that travel within the genome, replicating and altering the amount of genetic material. These transposons may result in mutations and alterations to the genome function in cells. Researchers have isolated certain transposons in the parasitic triatomine bug and discovered that 98% of them were identical to transposons found in animal hosts like certain opossums.

Small changes in the initial AI can lead to big impacts on the long run. Careful consideration will need to be made in terms of robustness, bias, reliability, etc. Since these tools are programmed by humans, the programmers may come with their inherited biases. To provide the best results, we need to ensure that the data used in developing the algorithms come from a diverse and bias-free source. That will help eliminate the bias – whether that bias is unconscious or conscious.

One problem is that proprietary algorithms are not being made available for auditing by independent entities. For example, a recent assessment of Google's translation software showed it translated gender-neutral pronouns from different languages into "she" when talking about a nurse and a "he" when talking about a doctor. Researchers at IBM are investigating and creating automated anti-bias processes when making decisions to inhibit biases. Constant checks for integrity will need to be undertaken to ensure long-term sustainability.

Laws of Nature and Ecosystems

In the long term, we believe the laws of nature and long-term sustainable equilibrium will prevail. We should be conscious of short-term dynamics that do not reflect real trends. We resemble nature, as individuals and as a society; society is a complex, adaptive, and interdependent ecosystem.

If society deviates from being an ecosystem, it may collapse. It is crucial, in times of fast disruptive technological change, that society adopts emerging technologies in a balanced way without cognitive bias. We should be wary of what may happen when we have too narrow a focus as opposed to taking a step back, observing, and evaluating the big picture on how all these upcoming technologies and their ecosystems will interact with each other and create new realities.

All emerging technologies are not inevitable or ineluctable: they are the products of human intellect, ingenuity, and ethical beliefs. We can and must embrace a responsible and deliberate approach to their development and diffusion across all strata of our societal fabric. In later chapters, we discuss in detail the mechanisms that may help us and in which we may wish to take part to influence the society that emerges from this period of upheaval.

Potential AI Effects in the Real World

Since AI sophistication and pervasiveness are continuously expanding on a continuum that we cannot fully grasp and anticipate, the key thresholds associated with non-linear evolution in capabilities are unclear at this stage. As the technology advances further and becomes more intertwined with the very fabric of our lives and with fundamental aspects of our society, the impacts could and will spread beyond privacy concerns or technological unemployment, with potentially devastating effects.

Since the beginning of the Industrial Revolution, productivity has surged across various industries. One of the greatest gains attributed to AI is that it delivers even more enhancements in productivity. Some people even consider AI as the ultimate tool for wealth creation. However, the increase in productivity in recent years has not been positively reflected on middle- or lower-class workers, and inequality continues to increase, the more advanced the technology becomes. Because automation is considered a substitute for labor capital, and more labor - whether blue or white collar - will continue to be displaced and substituted by advanced AI-driven automation, AI may be considered a driver that will augment inequality.

The automated processes created by AI will help users make better decisions. However, other challenges may arise when attempting to explain decisions made by an AI algorithm. You have probably encountered some innocuous – if annoying – examples online when you repeatedly receive advertisements for something you have just purchased. But some decisions being made by AI are more complex and harder to explain, like why a bank loan was rejected. For example, an innovative corporation called Smart Finance uses AI to assess over 5,000 personal features from all the data that are available to assess creditworthiness and risk assessment. By contrast, a conventional bank only assesses a fraction of those parameters. One of the parameters Smart Finance uses in assessing the application is by quantifying the history of the phone battery. According to them, a lower percentage of charge on a phone battery is an indication of a higher risk of defaulting on the loan. Another interesting parameter they use is how confident the user is while filling in the application process. Assessing and making decisions that affect humans may not always align with what AI deems important. Some people, like the authors of this book, for instance do not charge their batteries until they are almost empty as it is better for the long-term longevity of the battery. Why should we be denied a loan application just for looking out for our phone battery life?!

An additional aspect of AI to consider is its ability to collect tens of thousands of personal parameters from online users and mature behavioral prediction algorithms. These algorithms are capable of predicting a user's behavior and gauging certain personality traits to create personalized marketing content. This will create significant value for the advertising industry by minimizing uncertainty and focusing only on relevant content. What if all these advanced tools could be used in the wrong way to sway voters, for example? What impact would that have on democratic nations? In later chapters, we will examine in greater detail how and why this may have already happened, and what options we have to protect democratic processes.

Facial recognition technology enabled by AI has been the focus of heated debates and discussions, as more public and private entities move forward in implementing this technology. Facial recognition has become very effective with the use of advanced AI algorithms, and raises several civil rights and privacy concerns that need to be considered. The technology has the ability to control crime and protect the security of the workplace – but concerns for privacy violations, mass surveillance, and even reports of facial recognition technology failure when used on people of color are putting this technology in question. Regulations are still vague on what specific purposes the technology can be used for, and how it is to be used – including how long data can be stored. This will need to be properly addressed and regulated by lawmakers.

Technology to recognize an individual's features has also been extended to animals. A startup called Mooofarm in India claims its algorithm has 95% accuracy in distinguishing one cow from another and is used to fight insurance fraud. The algorithm requires pictures of each cow taken at different angles to identify unique features of the animal.

Recently, a number of startups have created systems enabling users to upload pictures into their system and match them to publicly available photos of that person by searching online assets including social media platforms. In the near future, this technology may even be available for consumers subscribing to companies' databases, which may contain millions of people's facial characteristics acquired from non-publicly available resources. What will be even more concerning is when this technology becomes commercially available on smartphones for anyone to use!

Is AI Really Going to Change Everything?

AI appears to polarize opinion. Why do some people assume that AI will either lead to a fiery apocalypse or a glorious utopia, and not something in between? We believe that these extreme scenarios are deeply rooted in how humans relate and emotionally react to change and uncertainty. We will endure as a society if AI capabilities and potentials are used properly – but that will take engagement from individuals. Put simply, you and your loved ones need to care about what happens.

A vivid example is the futurist idea (and related belief system) of the technological Singularity. In Ray Kurzweil's vision, the Singularity is the establishment of a sort of utopian paradise. Everyone is rendered immortal by biotechnology that can prevent and cure all illnesses; our brains will be uploaded to the cloud; inequality and suffering will disappear thanks to these technologies.

It's evident how the Singularity nurtures a similar narrative. As our technologies become more advanced and sophisticated, they increase our "control" over nature. If we pass through and emerge wiser from the Singularity and all of its threats and promises unscathed, then we will have found paradise. Most major technology disruption sounded apocalyptic when first conceived, and then it became commoditized and "domesticated." However, this absolutely does not guarantee that the same will happen to AI.

It seems like every major development provokes this kind of cautionary response. Nuclear weapons did it too, from the dark years of the Cold War up to today's debates: either using atomic power would prove the final straw and we would destroy ourselves, or the nuclear energy could be harnessed to build a better world. People talked at the dawn of the nuclear age about electricity that was "too cheap to meter." The scientists who worked on the bomb often thought that with such destructive power in human hands, we would be forced to cooperate and work together as a species. The debate continues to rage – with some countries choosing not to replace aging reactors and looking to other forms of renewable energy instead.

When we see the same response over and over again to different circumstances, cropping up in different areas, whether it's science, religion, or politics, we need to consider human biases. People tend to like millenarian beliefs, that we are counting down to some kind of judgment day or apocalypse. They somehow find comfort in joining a polarized end of the belief spectrum: according to these dynamics, once again, when the idea of artificial intelligence outstripping human intelligence emerges, these beliefs spring up around it. Most people don't base their decisions on facts and may not even analyze information – we are led by imagination, sometimes fear, and sometimes ideas from fiction. They (and We) are not as rational as we would like to think. We are creatures of narrative. Physicists observe the world and we weave our observations into narrative theories, stories about little billiard balls whizzing around and hitting each other, or space and time that bend and curve and expand. Historians try to make sense of an endless stream of events. We rely on stories in the desperate attempt to make sense of the past, justify the present, and prepare for the future.

We need to be wary of the millenarian narrative when we are considering technological developments and the Singularity -and existential risks in general. Maybe this time is different, but we have cried wolf many times before (although, of course, the key to that story is that in the end, a wolf comes). There is a more likely, less appealing story. Something along the lines of: there are many possibilities, none of them are inevitable, and lots of the outcomes are less extreme than we might think — or they might take far longer to unfold than expected. The problem with this more realistic and rational approach to reality is that it implies embracing a higher level of uncertainty. It's so much easier to think of the near future as either signaling the end of the world or the dawn of a utopia — or possibly both at once. It's a narrative we can get behind, a good story, and maybe, a nice dream.

But looked at more closely millenarian beliefs are not always the most promising ones, because they remove human agency from the equation. If you think that, say, the malicious use of algorithms, or the control of super intelligent AI, are serious and urgent problems that are worth solving, you can't be wedded to a belief system that insists utopia or dystopia are inevitable. You have to believe in the nuances or the complex reality we live in, and in your own ability as individuals as a society to influence the future we are creating - and take responsibility for your own life. As we move into an uncertain technological landscape, we need to be aware of the power and the limitations of dreams and we need to stop blindly believing what "the geniuses" tell us.

Substantial challenges and limitations still face AI in its current form. The big amounts of data generated everyday will continue to evolve and flourish, becoming the fuel that brings the AI engine to new capabilities. Ensuring bias-free algorithms and data used to train the AI is paramount. The security, robustness, ethics, accountability, liability, privacy, and transparency of the algorithms are also very important. The evolution and ubiquity of AI will require properly articulated regulations, policies, and ethical considerations as well. We hope that, in the regulators' priorities, concerns for AI's socio-economic effects in the real world will take precedence over semantic quibbles, and that the right moral considerations for society will eventually take priority and be adopted. What those right moral considerations will be, are in part for you to decide.

CHAPTER 2

APPLIED AI: INDUSTRIES REWIRED

Many traditional industries will be impacted by AI deployments in their core processes. Artificial intelligence enables computers to perform intelligent tasks that typically require human intelligence like visual perception, decision-making, and speech recognition. Although media hype suggests we should believe that advanced computers and autonomous, intelligent machines will take over the world in the near future, the reality is that over the past five decades attempts to replace human cognitive abilities has only resulted in modest gains in practice.

In most cases at the current stage of AI evolution, AI is synonymous with Machine Learning (ML) and predictive analytics with capabilities that have brought efficiency benefits and not the "Skynet-Singularity." The sense of gloom brought on by the hype -and the subsequent no doom-, has led many people to ignore the coming impacts AI will certainly bring. The difference today, compared with when AI was introduced in the 1950s, is that the most recent AI improvements and breakthroughs have begun advancing at a near exponential rate.

Since its origin in the 1950s, Artificial Intelligence has been the most divisive sub-discipline of Computer Science, raising animated debates. In the first years of the third millennium the discussion regained momentum when – thanks to the development of hardware and software technology – leading researchers and practitioners conceived novel theories and new theoretical and practical approaches to AI that reopened the debate, and the challenges associated with machines and human intelligence.

Instead of framing the debate as machines in competition with humans, we suggest considering computers and robots as rational helpers/supports to humans. Cognitive computing technology, with its superior capability of ordering data and information rationally, can help users to internalize rational criteria to augment their abilities to share contextual knowledge while facing challenging problems or novel situations.

> "I believe there is no deep difference between what can be achieved by a biological brain and what can be achieved by a computer. It, therefore, follows that computers can, in theory, emulate human intelligence — and exceed it."

From a speech given by Steven Hawking at the opening of the Leverhulme Centre of the Future of Intelligence, Cambridge, U.K., October 2016[2]

Human intelligence has dimensions like empathy, intuition, judgment, passion, emotions, etc., that current artificial intelligence cannot take into consideration. However, cognitive computing systems and AI can be a useful overlay to extend and/or augment the intelligence of humans and their inherent capacities to navigate complexity. In fact, computing systems have become definitively better, at the level of pure rationality, than human beings, as demonstrated by the popular examples of world champions of chess being defeated by computer algorithms. Another example is the Google AlphaGo AI that was able to defeat the Go world champion by using a strategy to maximize the effectiveness of each stone played individually. The board game Go originated in China over 3,000 years ago and opponents work to capture sections of the board using "stones." This complex game requires several layers of strategic thinking with around 10^{170} possible configurations in the board game.

Many of the recent successes that companies in various industries have showcased online and at tradeshows have in fact been related to implementations of AI in combination with large datasets also known as "Big Data." In many cases, AI has helped transform large volumes of structured and unstructured data into actionable knowledge. Whether it's consumer purchasing history or driving logs from "Waze" users and smart cars, data are used to instruct computers on how to do tasks that in the past could only be performed using intrinsic intelligence.

For example, many of us have phones or smart home assistant devices like Google, Alexa, Siri, or Bixby to help us with simple, repetitive, and predictable tasks. The current level of sophistication of these devices and their algorithms is low but it is starting to improve at a fast pace. Google recently released its AI Assistant that can make phone calls and conduct an intelligent conversation with humans to make appointments at restaurants, hair salons etc.

Samsung recently announced its first AI human simulation named NEON. NEON uses advanced AI algorithms combined with a virtual character that resembles and behaves like a human. The system is compatible with multiple languages and has the capacity to evolve and learn on a constant basis. Another AI human-like interface was recently created by Soul Machines using deep science and AI.

The emergence of the internet was an opportunity to use that base to create products and ecosystems that leveraged on that technology and has helped create several trillion-dollar economies. If we follow the same model, we can explore what the new "internet" technology is, and strategize about what products, industries, and companies are needed to serve these technologies. What will the ecosystems of all the emerging technologies be, and how will AI influence them? And how will we recognize them?

The profound reaction to viewing Earth from outside its atmosphere, referred to as the "overview effect", enabled some astronauts gazing back at planet Earth from space to experience an altered consciousness in which some felt a sense of connectedness with Earth and with one another. To better predict the state of future technologies like AI and their impact on society, we need to step back and observe the big picture from a distance similar to the overview effect. We need to question our perspectives at all times and understand the connections that exist across industries.

The Dream of Autonomous Transportation

We are all aware of the amazing evolution of autonomous cars. Humans have been dreaming about self-driving cars for as long as automobiles have been in existence. The first experiments in autonomous cars began in the 1920s, and real trials took place as early as the 1950s. But without the power of advanced technologies and Big Data, nothing succeeded. By the 1980s, the power of computers finally began to turn the dream into reality. Although autonomous cars in their current state of maturity have certainly not been without issues, it may not be long before they are commonplace on every road. How did we get here? The machine — in this case, the car — "learns."

Self-driving vehicles require challenging software development and a broad learning approach that enable vehicles to be exposed to a vast number of traffic situations. Driverless vehicles use a combination of advanced sensors and complex AI algorithms to respond to their surroundings. They depend on data collected by various sensors like cameras, radar systems, sonar, and laser illuminating detection and ranging (LIDAR). The LIDAR system bounces a laser beam off surrounding objects to construct a 3D map to identify possible hazards and their distance from the driving vehicle. An array of beams can cover up to 360 degrees view. The integrated radar system is used to monitor the velocity of surrounding objects in real time and the software reacts by reducing speed or maneuvering around objects. HD cameras can be used to recognize traffic lights and road signs. The sonar system, although it has a short effective range, can be used as a redundant system to verify data from other sensors.

The data from the sensors is passed to algorithms and classified accordingly. One of the main challenges is interpreting the image data and sensors properly. For example, distinguishing between another vehicle, a pedestrian, or a dog crossing the street. Sometimes, iterations are needed until all data are properly classified. As new images get fed into the algorithm, they will typically be classified with more precision and accuracy.

The vision-based safety system has been challenging due to the trade-off between fast image analysis and its accuracy. One of the challenges that have inhibited the computer vision algorithm is the lack of speed and accuracy in detecting pedestrians in images processed. When processing images with the intention of detecting real-time pedestrians, speed is a must. Researchers have been working on developing advanced algorithms that can perform these detections with high speed and minimum error with the eventual goal of being able to detect pedestrians in real time.

The process is surprisingly simple:

1. A human driver takes the machine for a ride. During the ride, the machine observes the vehicle's sensor readings and records each move the human driver makes. The machine then observes and records the outcome of those moves and

classifies the quality of the driver's decisions. Is the driver getting too close to the vehicle on the right? Is the driver getting too close to the curb lines? Is the driver too close to an object ahead? Are the car's sensors giving the driver warning signals?

2. The machine records all observations and outcomes over time. The machine builds a database of information using every observation, every reading of the vehicle's sensors and every decision by the driver.

3. The machine generalizes the data to make intelligent decisions in new situations. Based on all past situations, the machine can now make highly intelligent decisions — even when it encounters new situations on the road that it has not observed previously.

4. Over time, the machine gets smarter. The more data that are available, the more inclusive the machine can be — essentially learning from itself.

Yet, as remarkable as autonomous vehicles are, the technology is actually quite simple. Why? Because there is a finite number of decisions a driver can make in any given situation. As a result, the car uses data-driven analysis to detect good or bad behavior and adapts accordingly, becoming "smart" in the process. By combining that simple process with big data, autonomous cars may be on the cusp of turning a century-old dream into reality.

So, what does the future look like with autonomous cars?

First, let us look at the implications for driving safety and the reduced number of accidents. According to the National Safety Council, it is estimated that tens of thousands of people are killed in the US alone, and several millions injured due to traffic accidents. One of the main goals for autonomous cars is to achieve zero accidents by taking out the weakest link in the driving process: human error. If successfully implemented and the technology is matured, that alone can result in significant saving of lives and eliminating the large majority of injuries on the road.

Second, what influence can autonomous cars have on increased productivity? According to a Harvard University study, the average American spends 100 minutes per day driving. That is almost 40,000 hours of driving in a lifetime. Furthermore, motor vehicle crashes in the US alone can have a big economic toll every year. Therefore, an increase in productivity due to reduced accidents, ability to work while on the road, etc. could be an important catalyst for economic growth.

Third, what is the impact from autonomous cars on traffic and the environment? Imagine a car that can drop children at school, pick up the parents from home, and drive them to work, and then be on standby for the next pickup. Autonomous cars will significantly contribute to the reduction in the number of cars on the road. Families will no longer need two, three, or four cars. One single vehicle will do that job. And imagine the time and space saved on parking! In addition, interactions and communications among autonomous cars will enable cars to be at a closer distance while driving at high speeds, thus reducing or eliminating traffic congestion. The reduction in the number of cars will create an environment with significantly less traffic, which translates to less pollution and more environmentally friendly towns across the world. According to a Yale University study, there is a fleet of 246 million vehicles in the United States alone, making them a potent polluting force in bulk. That number is geared to be significantly reduced once these autonomous cars become mainstream.

Fourth, what about the impact on the older population? Aging demographics won't have to be concerned about having their driver's licenses revoked due to old age. All their transportation needs can be fulfilled through this technology. There are numerous benefits from embracing a near future where autonomous cars will be part of everyday life, but do all these advantages come without a cost?

There will probably be little impact on current young generations that are growing with these technologies as part of their lives, but what about in the near term? How would a person that has been driving his entire life feel about being transported in a vehicle where he has no control over it, driven at 80 miles per hour on a freeway? That is not going to be easy to get accustomed to.

Driving on a daily basis entails the use of several cognitive skills as we navigate through various driving scenarios. The older we get, the more we tend to exhibit a decline in those cognitive abilities. Some recent studies have indicated that driving may actually help reduce age-related cognitive decline. We also will need to research long-term cognitive decline from lack of driving as more driverless cars make their way to the streets.

What are the long-term safety implications? One of the main components guiding these driverless cars employs laser technology. What would the long-term impact be on humans from exposure to thousands of cars passing by while radiating these energies? Further research is needed to ascertain any future health effects and how to help mitigate them effectively.

As passengers are expected to spend extended periods of time in these vehicles, new technologies that provide services like a mobile office, advanced entertainment systems, food service systems, etc. will emerge creating new economies and business models. The potential adverse implications on insurance companies, car manufacturers, commercial parking spaces, and the energy sector will need to be considered by these different industries. Executives from these industries will need to adapt and transform their existing value proposition to remain relevant in this wave of emerging technology.

What about autonomous trucks?

"I just realized that with the advent of self-driving vehicles, we may soon get country songs where your truck leaves you too." Marko Kloos, Twitter

Another example to explore is the trucking industry, which is the lifeblood of the US economy. The industry is responsible for most freight transportation over land. Over 70% of all the freight moved in the US is moved on trucks with more than 3.5 million truck drivers using billions of gallons of fuel. The trucking industry expects many drivers to retire in the next two decades and has been struggling to replenish the aging population of drivers with new ones to join the industry. Trucks represent about 5% of the vehicle population in the United States but consume approximately 20% of total transportation fuel. The American Trucking Association indicates that the trucking industry spent $105.2 billion buying diesel fuel in 2015. Trials of experimental driverless trucks have already been taking place in the US. It is expected this technology will increase its penetration of the industry with further changes to take place in the next decade.

Truck driving is a hard job with countless hours driving behind the wheel. According to the Federal Motor Carrier Safety Administration, hundreds of thousands of trucking accidents occurred every year in the United States with thousands of fatal crashes, and tens of thousands of crashes causing injuries. Distractions and exhaustion due to long hours of driving are among the key safety issues we face today and some of the main contributors to accidents. Driver assist and collision avoidance technologies have been making headway in the automotive industry, contributing to the reduction of accidents. This technology will be replaced incrementally with autonomous driving systems.

Initially, long-haul trucking routes will have segments that will be completely automated, and others will involve a human in the loop. This combined system will eventually be replaced with complete autonomy across the whole route. Advanced technologies like AI and spatial sensing technologies will contain complex systems and algorithms that can make the optimal choices to improve safety. Tesla unveiled a new prototype for a battery-powered, almost self-driving semi-truck. The new prototype is supposed to be more efficient and cheaper to run than conventional trucks. Recently, a company named Embark has been pushing the state of the art in autonomous driving and has developed advanced technology for use in autonomous trucks. That technology is already being used to move freight for several companies in the southwest US. The technology is even capable of autonomously driving and navigating trucks in various adverse weather including rainy and foggy conditions. That technology will continue to expand until it is able to reach more locations in different driving conditions including during severe weather.

Another important component that needs to be analyzed and modeled is the behavior dynamics of the vehicle's surroundings. These are needed to allow the self-driving technology to respond properly to certain situations. For example, a vehicle driving at a low speed on the left lane implies a high probability that the car following will attempt to pass, or a car driving erratically may suggest surrounding cars slowing down; a cyclist on the right lane may also imply that cars will slow down and attempt to change lanes. With AI advancements, it will be possible for systems to perform several simultaneous tasks faster than humans can. If successfully implemented and the technology is matured, that alone can result in significant reduction in loss of human life and eliminate the large majority of road injuries.

A positioning system is essential for the navigation of driverless vehicles. The structure of the maps needs to be defined with more details than the conventional street maps currently available from Google or Apple. Mapping the proper area for autonomous driving and keeping the maps updated is a big challenge, especially when done for the whole of the US or internationally. Trucking companies can initially limit the routes to areas that have been properly mapped with high fidelity. The routes can then expand as the mapping continues to expand to broader areas.

Having autonomous cars with virtually no accidents will reduce insurance premiums substantially. AI can deliver powerful predictive analytics that will reduce the maintenance cost of vehicles by selecting the best time to replace components that are about to fail – not only reducing the maintenance cost but also preventing breakdowns. Artificial Intelligence and Big Data will be employed to analyze the optimal routes and driving conditions like weather, traffic, road conditions, etc. Acceleration and deceleration will make a difference on the amount of fuel consumed. Since autonomous driving trucks will have the capacity to communicate and interact with each other, truck convoys can drive closer to each other at faster speeds. That will minimize the constant acceleration and deceleration that normally result in increased fuel consumption. Also, the reduced distance between trucks will minimize drag and thus increase efficiency. AI algorithms will help optimize these driving scenarios even further.

Another question that comes to mind is the ethics of the complex algorithms and software that will guide these advanced trucks? Evoking the traditional "trolley problem," for example, how will a truck react if it had to choose between impacting a car with a toddler inside and hitting an older person on the side of the road? Would it give preference to the driver? There are multiple complex scenarios that may be construed as controversial but will eventually need to be identified and resolved. Will humans have the capability to override a decision that the autonomous trucks have made?

Looking at the aviation industry, modern airplanes have been able to fly autonomously for some time. The advanced navigation systems are capable of warning pilots about potential concerns; however, most of the accidents in recent years have been attributed to pilot error irrespective of the fact that the system displayed signs indicating problems were occurring. Will the same behavior translate to autonomous trucks and human drivers, or will the industry be more receptive than we have been toward airplanes?

As we have outlined, autonomous vehicles employ a variety of advanced systems and algorithms to enable their autonomy and ascertain their surroundings. Typically, these vehicles are connected to the manufacturer to perform system updates and monitor their performance. As discussed, they will also be connected to other driving trucks to optimize the driving experience. Could these connections constitute a weakness? The potential of hacking an autonomous driving truck and exploiting vulnerabilities could lead to new challenges and risks and may even result in physical danger. These cyberattacks could result in substantial disruption to the trucking industry and the economy. Artificial Intelligence and Machine Learning will play an essential role for companies looking to keep their resources secure. These technologies will add several layers of defense to protect the trucks and drivers. The government must play an important role in establishing legislation that will become the standard for commercial companies to abide by. This area is starting to attract a variety of new startups and investment opportunities.

Big Data collected from various sources, including other trucks driving on the same route, will help regulate speed and optimize routes. The AI can be very helpful to trucking companies in predicting demand; this ability could translate into better optimized and enhanced logistics and faster deliveries. All of this will enhance risk management and increase profitability. The introduction of autonomous trucks will potentially free millions of drivers to perform other jobs that may be difficult to replace with autonomy and AI, so they could then continue to contribute to the economy. The fact that an autonomous vehicle doesn't need to stop and rest means that products can be transported across the nation much faster and at lower cost. The rise of AI and autonomous trucks will also open the door to other adjacent and complementary industries that may serve that market.

AI in the Construction Industry

Several industries have started adopting AI to improve various roles in their sector like productivity, efficiency, safety etc. This integration will continue to move forward and will become a cornerstone of many industries. The construction field is no exception and is considered one of the industry sectors to benefit greatly from advances in AI, including improving productivity, risk and schedule management, and safety. Most projects in construction are unique and AI will serve to assist humans in dealing with the many external factors and dynamic environments.

Although it offers great benefits, the adoption of AI in the construction field will face several challenges. In general, the construction industry has been lagging in implementing new technologies compared with others. Some of the limiting factors to adoption include the perceived increase in risk from changing conventional methods, the large capital investment needed upfront, and the training required to manage the new technology environment. Early adopters will have an edge over their competition. Overall, the increase in productivity from using Artificial Intelligence in the construction industry is overwhelming and eventually that sector will harness its power, instead of resisting, to gain enormous benefits.

To further assess the impact of AI on various phases of construction, we will evaluate each of the three segments.

In the **pre-construction phase**, surveys are an important component. Advances in drone technologies are now allowing machine vision to be used for surveys. These technologies are significantly reducing the time needed to compose accurate survey maps. Similar technologies can also be used to construct 3D maps and other plans needed before construction begins. These advances will result in meaningful savings in labor and financial resources. Other important uses for AI include optimizing complex project schedules, considering the large number of possible alternatives and enhancing project planning and better on-time completion of tasks.

One of the advantages of AI is the ability to learn from previous data to improve future performance. That methodology can be used to assess previous contract bids, optimize the bid award rate, and improve its margins. The same technology will help engineers and architects make better decisions and create better designs in less time. AI can also provide better insight when critical decisions are made by estimating the risks of projects, which otherwise will be difficult to ascertain. The construction industry can take advantage of innovations and advances in AI across various industries and implement these tools in the construction field. For example, they could easily take advantage of the technology that the retail industry is using to optimize the supply chain to reduce oversupply and cost, and adapt it for the construction industry.

In the **construction phase**, image recognition technologies used in several industries can be adapted and combined with imagery from drones to implement quality control, track progress, and compare construction discrepancies that deviate from the existing designs and drawings. All of that can be performed without being physically on the job site. The image recognition collected from drones can also evaluate data collected on work sites to identify unsafe workers and use this information for future corrective action and training. The altitude and different perspective acquired using drones can help project managers identify issues and flag unsafe behaviors during the construction process.

Reports from project managers indicate a challenge in recruiting new workers to fill the industry's need. Advanced robots have been used to replace several manual tasks and save time and potential injuries sustained by workers on construction sites. Literature has shown that some robots were capable of laying blocks at a significantly higher rate than humans were. Advances in autonomous vehicles in other industries can be transferred to construction equipment and machinery, and used to help construction sites with drilling, transportation, and hauling. AI can be very helpful in predicting demand that could translate into better optimized and enhanced logistics and faster deliverables and on-demand delivery. All this will enhance risk management and increase profitability.

Once **construction is complete**, technology within buildings can make them better for their users and the environment. The IoT (Internet of Things) is automating homes to improve energy effectiveness and make our homes more efficient and safer. That technology is constantly improving and will soon be an essential part of any residential or commercial building due to the many practical and economic benefits it offers to builders and users. Technologies or solutions developed from adjacent industries that once were specific and restrictive to certain fields, are starting to find new applications across industries, as an ecosystem. For example, AI can collect and analyze data from various sensors to understand and determine patterns that can be used to deploy real-time solutions, prioritize preventative maintenance, and reduce cost in buildings just as it can in vehicles.

AI in Education

AI is transcending and disrupting industries across the board by transforming and challenging conventional methods. AI has an enormous role to play in the evolution of the educational system and the educational experience of the future. The ability to sift through the complexities of interpreting and processing enormous amounts of unstructured data gives AI an important role in modifying and revolutionizing the educational system. The fact that we are at the narrow phase of AI right now should not anchor us from the enormous potential AI will have once it has reached the General AI phase. AI has the ability to drastically change the educational system as we know it today and transform how the learning content is delivered. The future educational experience may even result in a more personalized and adaptive experience that aligns with the required learning outcomes.

Technology is advancing at an exponential rate, and education, and learning and development programs should continue to be updated to reflect that fact. As humans, our intellect and learning capabilities are different from one person to another. Education is headed into a personalized educational experience using the latest algorithms, deep learning, and tailored learning processes. Smart data mining and gathering systems that are enabled by advanced computing systems will also help substantially.

If we look at the conventional classroom from centuries ago, the standard is to have a teacher in front of the class providing knowledge to a number of students in the class at the same level, not taking into account the variability in each student's intellect. Some students will be left behind struggling to understand the concepts, while others will be ahead of the information being taught. In essence, the teachers teach and deliver the content to the average of the class. Students' needs are different: some students learn by repetition, while others understand by examples, etc. In order to deliver the best educational experience, the delivery of content has to be personalized according to the student's intellectual ability and learning behavior in a way that is adaptive to the student's needs. In reality, it is not possible to do so with the current educational system due to the lack of resources. The new learning process also needs to be complemented with social activities outside the individualized learning process in order to give a rounded approach to a better education.

As technology is advancing at a fast rate, so should the curriculum used in teaching. By the time a five-year engineering student graduates from college, some of the information learned may be obsolete and not relevant anymore. AI can help by continuously working to ensure the knowledge taught is up to date, and by regularly reassessing changes needed to the curriculum and incorporating them as they occur. AI will also be able to provide better comprehension gathered and analyzed from large sets of data.

The system may help the students learn at their own pace until they master certain topics or skills. That in return will give better insights into the learner's behavior and will improve the customized learning program based on the needs. It will also help understand the learning behavior of the students and better understand how students learn and absorb information. Using all these methods AI will be better able to track the students' progress, performance, and retention, and then adapt the system as needed to become more effective. Another important element that needs to be incorporated into the learning process will be developing AI algorithms that help students learn independent thinking and creativity.

Bias is also an important element in the educational experience; in certain cases, a teacher may be biased in favor of or against a student, which may be reflected in the grading of his or her homework, exams, projects, etc. The same can occur when developing the AI tools: we need to ensure that we account for bias and eliminate it, especially when it comes to interpreting the analysis of learning progress, or what material and curriculum better fits different individuals.

AI learning systems may also help physiologically as some students, for example, feel uncomfortable when being put on the spot in front of their classmates. Artificial intelligence will enable students to learn in a judgment-free environment. AI systems can offer suggestions for improvement on a private basis making certain personality types more at ease. In the short term, however, AI-based individual education may fall short of providing assessments compared with what a human teacher can do, for instance having an eye for "character," "integrity," and an understanding of the unique individual's characteristics and personality. We expect these systems will augment what human educators are doing in the short term, but eventually may become more independent and replace certain/most educators' tasks as the technology ecosystem continues to mature.

An automated type of learning may be good at the personalized educational content, but risky in the short term when determining anything else in individual development. As the technology matures in the upcoming decades, this statement will become less relevant. We anticipate that this technology will also result in a reduction in the total cost of education and may bring new tools to help students with disabilities to better adapt and learn. Will these technologically advanced tools level up the playing field in terms of learning experiences? Students using AI systems can learn from anywhere in the world at any time. If these tools are made ubiquitous for students around the world including in poor and developing countries, would that help improve their long-term education? Would it result in superior economic outcomes and societies?

AI in the Healthcare Industry

Artificial Intelligence is expected to have an enormous impact on healthcare in terms of quality, efficiency, and performance. Many healthcare institutes have already started to harvest the value from AI-integrated systems, and more adoption is expected to continue at various levels. Exploring AI's impacts on the health and medical sector would require several books, but this section will touch on some of the essential points.

Recent advances in data science have significantly expanded and augmented the potential of AI and ML for several medical applications. AI is a new power tool in the development of medications. Researchers can use AI in the quest for the exploration of new indications aimed at existing drugs and gene therapy. From distributed clinical trials that bring trials to patients, to computational AI-based frameworks that match chemical identity to genetic mechanisms of action, AI presents many opportunities to accelerate the discovery of new drugs, to develop more efficient processes, and to predict multi-drug interactions and side effects. Fundamental innovations in biotechnology and biomedicine are identifying and implementing algorithms of relevance for developing novel drugs; this will make personalized medicine and individualized therapeutic treatment a reality. For example, AI researchers at the Massachusetts Institute of Technology have figured out how to leverage AI to predict novel forms of antibiotics by training it to predict molecules with bacteria-killing properties. The research has resulted in the discovery of a new compound that works on strains that had been resistant to drugs.

AI systems will have the ability to take advantage of unsupervised learning and allow deep learning algorithms to figure out trends independently, beyond the ability of humans. This technology will allow healthcare professionals to perform patient assessments more accurately, provide faster diagnoses of diseases, and possibly improve early detection of life-threatening conditions, prompting timely medical interventions and more effective treatment options. We can leverage deep neural networks to learn from large repositories of images.

Similar real-time, interpretive analytic tools, which are also used in autonomous systems, can help us understand the progression of many severe medical conditions. Examples include monitoring acute macular degeneration, quantifying fatigue attributable to multiple sclerosis, and measuring the behavioral manifestations of genetic autism disorders in preclinical models.

Advances in AI and quantum computing will be instrumental in the emergence of nanobots, which have the potential to help fight diseases in the upcoming decades. AI can minimize the level of error in the geometry of the particles, optimizing the nanobots' development and accelerating their manufacture. Combined with nanomedicine, nanobots could be the key for preventing and curing chronic diseases at the molecular level. Nanobots could also be instrumental in repairing cells, correcting genetic mutations, and boosting immunotherapy.

Despite the huge potential of advanced AI to transform data analysis, the real breakthrough from a social and economic viewpoint would be empathic AI. This AI would be able to take complex decisions under uncertainty and leveraging multiple intelligences – like humans – in order to transform at scale the economics of healthcare, patient care, elderly care, and childcare, with reliable, technologically trustworthy, and actionable solutions, affordable for all.

Medical scanning systems are also starting to incorporate AI platforms in order to generate better image quality and improved characterization of organs. AI algorithms continue to be updated and improved to detect specific diseases by better analysis of scanned images.

The incorporation of AI systems has the potential to reduce the amount of radiation to which patients are exposed in most current medical practices, combined with the desirable capability of improved image-recognition and interpretation, with better accuracy and speed.

For example, a study examined an AI tool designed to detect lung cancer tumors from CT scans. The evaluations obtained were then compared to results characterized from several radiologists. The outcomes indicated that the AI system was better able to predict the result when prior CT scans were not available for comparisons. Similar results were obtained when an AI system was able to outperform a number of Stanford radiologists. Another recent study conducted by Google presented an AI model with the capacity to identify cancer in mammograms with more accuracy than radiologists including fewer false positives and false negatives. In the current stage of AI development, however, it is reasonable to assert that a combination of radiologists and AI-enabled systems will produce better results when compared with either an AI or a radiologist functioning independently.

AI technology has also shown the potential to predict irregular heart rhythm before it manifests in patients. Another example by Verily (the life science division of Alphabet) showed promising AI and machine learning in healthcare. The system demonstrated the capacity to screen for diabetic retinopathy and macular edema for patients. This would be an important tool and would offer great benefits for regions where doctors are scarce.

The year 2020 marks 20 years of continuous habitation on the International Space Station and the US now has its sights set firmly on returning to the Moon and putting boots on Mars. Being confined in limited spaces for extended durations could take a toll on the mental and emotional state of astronauts. AI is being considered for the well-being of astronauts during long-duration space exploration missions, for example, to evaluate and identify emotions by evaluating facial expressions, voice, and body language. The results and attributes can then be characterized and correlated to certain environmental and social settings. The system will be able to respond with empathy, mitigating the emotional challenges of astronauts. AI has been also found useful helping astronauts observing their health, adapting their nutrition, and fitness training to minimize muscle loss in the microgravity environment of space.

One of the expected challenges is the integration of AI systems into existing medical infrastructure. However, this challenge is expected to become less of a roadblock as more healthcare facilities adopt AI systems. AI investment in healthcare will continue to increase and show value in refining diagnostic abilities while reducing clinical errors and improving patient flow optimization.

AI impact on Creative Industries

Art and creativity are defining characteristics of being human. Is there a place for AI in this domain? Emerging technologies like 3D printing, holographic photonics, AI and ML applied to image recognition, etc. have the potential to augment artists' capabilities and output by providing unique creative options as the result of the aggregation and almost real-time analysis of vast sets of data and complex information from various sources.

AI and ML in the Music Industry

The music app Shazam, acquired in September 2018 by Apple Inc. for a reported $400 million, pioneered the space of music identification with its machine learning advanced pattern-recognition that allows users to identify songs they are listening to.

AI and ML-driven music generation algorithms have been used by experimental digital music artists to collect, analyze, index and cluster millions of songs, music reviews, music-related articles, and user-generated content on social media over a period of five years. The artists used them to identify revealing correlations between highly acclaimed, top-chart scoring songs and specific features or recurrent elements found in successful music tracks, in terms of patterns in pitches (or notes), rhythm, time and meter (the ways time is organized in a piece of music), three main melodic characteristics (phrase, contour, interval), harmony, and keys.

Early evidence showed encouraging results, with AI-inspired songs reaching top ranked positions in the iTunes Hot Tracks chart within 48 hours of release.

Of course, commercial success and mass-market appreciation are not necessarily indicators of music quality, originality, nor innovativeness; on the contrary, the very reason of widespread success of many music tracks is their familiarity or the fact that they are "unmistakably reminiscent" of previously successful songs.

Musicians can now use AI and ML to better understand the preferences of their audience, accurately predict which song(s) will be successful, tailor songs to specific customer segments, and help shift the creative process into novel domains. These technological capabilities have lowered the bar to success for many musical artists and have created quick profit opportunities for the media industry by scientifically designing and building pop stars, even in the absence of music talent.

AI-Enabled Design of TV Ads

In May 2017, media agency McCann claimed they had created the world's first AI-designed TV commercial, with apparently comparable results to human-directed TV commercials in terms of the target audience's appreciation. The *modus operandi* followed to achieve this result was basically to assemble a taxonomy database of deconstructed elements of TV ads (target audience, value proposition, music genre, context, overall tone of the communication, key message and related claims, etc.) extracted from all the award-winning TV ads in Japan over the previous decade. Visual recognition, natural language processing, and color-matching tools were then used to identify objects and themes. Each deconstructed element was then mapped and tagged according to multiple "dimensions" that were supposed to be correlated with the success of the TV ad.

As a second operational step, the database was probed and filtered by using a creative brief self-defined by a client, with a subset of elements they wanted to have in the TV ad. The AI algorithms were able to quickly and effectively "sift through" the database and extract suitable elements that best fit with the client-specified criteria. Humans produced the actual TV ad, but AI was a key element of the selection process. At the end of the "creative" process, a nationwide poll was conducted to assess consumers' preferences between the "AI-enabled TV ad" and a TV ad produced by humans in the traditional way. The two TV ads scored similarly; the AI-driven one was largely preferred by advertising executives (representing an audience of experts) while 54% of the target consumers preferred the human-created ad.

This example confirms the potential for AI-enabled searching through large datasets, and at the same time the current limitations in extracting real meaning, making sense of disconnected elements, and ultimately producing a cohesive narrative, for which humans are still needed. The key aspect of the intrinsic current limitations of AI and ML is the reliance on historic datasets and prior knowledge as *learning sources.*

It's clear that humans' creative processes can be augmented and accelerated by AI and ML by allowing the contextual access to potentially thousands of sources of inspiration that were produced over many centuries. This may yield particularly effective results in democratizing creativity and artistic production, dramatically reducing the amount of time and skill necessary to produce a good quality result.

The real question is what AI and ML will be able do in uncharted territories, when there are no datasets to rely on, when there are no previous references to "learn from." The uncharted territory is still by far the domain in which we humans give our best, make sense of various types of complexity, are able to identify emerging patterns, and ultimately find original pathways to create novelty and new knowledge.

Human intellect, ignited by human ambition and the innate drive to overcome limits, remains the main source of discovery and creation of novelty in our world.

By stimulating their intellect and by feeding their human passions, humans learn the necessary tools to create. It's their intellect that drives the creation of novelty.

AI-enabled Fashion Designer

Apparel design and fashion are a fertile ground for using emerging technologies to take the industry to the next level. AI and machine learning are already transforming sales using predictive analytics and improving the user's experience. By gathering more data about the customers, the industry can gain a deeper understanding of the customer's interests and buying patterns and can suggest bespoke customized products based on the individual interests. AI will also help predict trends in the market and generate products that are in line with the upcoming trends with much better accuracy. AI can also better manage logistics and the supply chain.

AI technology can also be combined with other emerging technologies like 3D printing. Customers will soon be able to personalize and order printed garments and be able to buy files and print clothes at home or at designated stores. 3D printing is being explored to engineer substantially better knits resulting in a more comfortable wear, as the 3D printed knit garments will contour and move better with the body due to being engineered in multiple dimensions. Since this printing process creates almost no waste, it will be better for the environment as well. The 3D printing technologies will create products at a faster speed and turnaround since steps like laying and cutting out a pattern can be skipped. 3D print-knits are also completely seamless since the shape is programmed. That process will make them last longer than traditional garments that are vulnerable at the seams.

We can envision having an AI-driven personal designer based on each individual, fashion style, personality, market trend etc. That design will then be produced through innovative 3D printing systems for use.

Artificial intelligence may be deployed in potentially promising ways in other creative sectors including music (the pop music already discussed and a range of other genres), art, paint, poetry, and culinary art. In the culinary arts, for example, AI can be used for suggesting novel ingredient combinations, creative or unique dishes that human chefs would never consider. AI and ML can augment and accelerate human abilities to navigate and analyze traditional recipes, food-related data, nutritional information, and physical and chemical properties of specific ingredients, in order to identify novel combinations and invent new dishes.

These are but a few examples. AI will transform the nature of our society in ways never experienced before. Machines will begin to perform tasks commonly done by humans, and eventually go beyond what humans can do. What will society look like in the coming decades? Will AI result in a utopian or dystopian society? We hope what we present in these pages help inform and direct those changes to an acceptable human result. Only time will tell!

CHAPTER 3

AI – IN WHOSE IMAGE?

In Principio

As we have hinted at it, Artificial Intelligence has the potential to transform not only the economy and society, but its promise of power creates wedges that may well become ideological and dogmatic due to the key interests associated. Simply put, there is too much at stake: this aspect is best demonstrated with an anecdote.

In the summer of 2018, when the outlines of this book first started to coalesce, I [3]paid one of my regular visits to an acquaintance of mine, a senior executive at a major tech financial firm in Silicon Valley. We were chatting cordially on the top floor of a skyscraper with the azure of the Bay surrounding us. During the catch-up phase, I found myself blurting out that we had started working on a book dedicated to investigating the transformational power of AI in both the present and the near future. As a tech insider, he gave a knowing nod but also flinched as if stung by an imaginary bee. "I suppose you do not intend to write anything negative, do you? " he asked with a half-smile. It could have been a half-joke, but there was a glint of steel in his eyes that made me pause.

I responded by saying that our author team had defined our mission as "making sense of Al": sort of a rebalancing act for curious, well-read people worldwide. Our prospective readers may not be AI experts; nevertheless, they were the kind of people who would feel the need not to become victims of this unprecedented wave of transformation. Our feeling was that they simply wanted to understand the basics that matter, eager to stay informed on what is already impacting them and what more would be coming soon, to get properly prepared. Their current options of getting unbiased knowledge are very limited and often distorted, we thought.

He was still uneasy and fidgety. "Let me tell you a story," he finally uttered after a long pause. "Not that long ago we published an AI study as part of a series we do periodically on frontier technologies. Responsibility for the study landed on the plate of the X team." (X stands for a nation known for its unbiased, analytical, uncompromising approach.) "They did a thorough job, showing a balanced, in-depth perspective about opportunities and risks. Regardless of its high quality, it felt, well – different. We at corporate were a bit concerned but ended up publishing it without any changes."

"The next two weeks were, frankly, a nightmare. The phone lines were burning with most of our clients calling our senior executives day and night, who were sweating like an ox. 'Are you ******* trying to bankrupt us? Have you guys all gone insane?' Yadda, yadda, yadda. So, we had to apologize profusely and make a conscious business decision to paint a rosier picture going forward and thereby keep our jobs," he said with a sigh.

Despite his kind, not-too-veiled intentions to warn me, I left that meeting reinforced in my conviction that this book needs to be written, and it needs to be written *sine ira et studio*. The tech financier probably has no idea how much he helped me overcome my early hesitation. Before that conversation, I had been semi-paralyzed by the "impostor effect", I guess – I had read many books by AI research experts whose combined knowledge seemed overpowering, but obscure for the public. The more voices that felt the need to join the chorus, the more apparent the disconnect between bold statements and the public's understanding became.

We as authors realized we needed to cross massive woodlands of obfuscation: nevertheless, the impact magnitude associated with the coming of AI was largely apparent to us from the outset. Also, at that time, there were certain kinds of important voices that were missing from mainstream AI literature: media were pursuing their next clickbait headline, closely followed by the entertainment industry; politics was just waking up to the phenomenon; religion was silent; social impact organizations were still trying to wrap their heads around the possibilities ... It was – and to a great extent, still is – a time of confusion, curiosity and concern, all mixed up.

So we dove in, and spent many months in confidential conversations with senior stakeholders around the world. We talked to politicians and government officials, who now emphasize the importance of AI every day to the press and their constituents, trying to bridge a growing chasm between AI and policy. We listened to corporate leaders who are betting big on AI and to their consultants who sell it to them. We consulted AI researchers, many of whom are finally leaving the ivory tower behind and signing up to enlightening the public. We talked to startups who expect elevated exposure and ample funding if they label themselves as AI (and some who are walking the talk rather impressively) and to impact leaders, who are playing catch-up with the relentless ambition and pace of their AI-obsessed donors.

The most common phenomenon we observed in this AI pilgrimage of ours was the discrepancy between the attitudes of these world leaders when they spoke from the pulpit, and when they spoke off the record, in tight circles they trust. Under limited pairs of eyes, rambunctious optimists become concerned leaders, users, consumers, citizens, and parents, when it comes to the threats and changes that AI will bring. The farther you go from Silicon Valley, the stronger this pattern becomes.

This psychological phenomenon fascinated us enough to dig deeper: what kind of power is capable of creating this level of "split personality": what is its nature? At this point, it became clear to us that without transparent truth-seeking and sense-making efforts, our society would stay trapped in these myths of inevitability or within the organizational logic that we are asked to serve, irrespective of what good people and critical thinkers we are. Almost unnoticed, we are relinquishing our intellectual freedom and potential to act – all at a moment in time that, so far, looks like the biggest turning point in human history.

On this journey of discovery, the question we kept getting was whether we were AI optimists or pessimists – and we find this framing most disturbing, especially when it precedes understanding. Why should it be one or the other? It is a sorry state of the public mind – and a consequence of the very destructive binarism permeating our contemporary societies – when a topic as complex and undecipherable as AI becomes a modern-day "homousion/homouision". This binarism – probably originating from the culture warfare trenches sadly dominating this decade - pressures us to label ourselves as AI optimists or pessimists, in the same vein as you might be expected to support a sports team (see Pittsburgh Pessimists vs. Oklahoma Optimists). These labels never made sense to us – they are actually harmful to all kinds of honest and balanced analysis. Our principle should be not to adopt AI-related prophecies mindlessly, and therefore we set out investigate major present trends and scenarios for what they are. When any of us sees emerging patterns that call for a warning – moreover, concerted action – there should be no hesitation in calling them out. When we see under-use of AI for good, or find more emerging use cases for trustworthy and beneficial AI, we lead the cheer. Ultimately, – if pressed-, the authors prefer to define ourselves as "equally thrilled and concerned about AI."

But polarization in the topic is very media-friendly, especially social media-friendly. Sadly, it also extends beyond the media, to more respected domains of research and reporting. One of the most comprehensive AI studies we came across polled AI experts' assessments in this very spirit of "yay or nay" binarism: the polarization pattern was even more visible and disconcerting. In the study, the biggest chunk of the experts' narrative analysis was mostly red flags and deep criticism on the unwanted consequences of the AI Shift, but by contrast the executive summary surprisingly found that two thirds of the same experts were by and large pro-AI: AI optimists if you will. That convinced us that – especially if the expectations of a technology wave have the capability of distorting the results of prestigious studies – this psychological phenomenon of self-censorship also warrants further digging.

In brief, there are risks associated not only with AI failing to operate as intended by humans on both the programming and receiving end; but also with the potential abuse of the vast powers of AI by those who invest heavily into it. In both sci-fi and professional AI literature, there are long-term hypothetical risks of AI developing full autonomy, breaking free and running amok. This latter danger may well be far in the future but we cannot fully ignore it: the other two are clear and present now, therefore deserve our primary scrutiny for this book.

Most AI authors talk about AI in a future sense: it is all too natural that both AI researchers and the sci-fi-loving general population prefer it that way. But a very good morning to you all: AI is here. So, as a contrast to that trend, we decided that instead of future tense, we will primarily employ "present imperfect continuous," meaning focus on what already exists "live" in production. The most intriguing part was the discovery of how these technologies and the impact they create *today* contribute to emerging transformational trends and to our emerging AI future: we are going to investigate these various scenarios now.

Into The Labyrinth

For the first time ever, the circular and the incremental forces of history have been upset by the open-ended exponential. Of these exponential forces, AI dwarfs all others in its transformational potential. We believe that due to the proximity of machines that think and learn, many rules will be rewritten - and be rewritten continuously.

AI has landed during a human age of mass disruption, fragmentation, disarray, and ethical rudderlessness. While this aspect is still not widely debated in public, we believe that recent political upheavals were considerably due to that explosive concoction of rising insecurity, inequality, and meaninglessness, that are, by and large, a byproduct of the digital realm taking over. Much more directly than the opaque effects of the digital, another factor and ugly truth is the subtle rise of job automation that has been emerging in several industries since the mid-90s – but accelerating exponentially and expanding since then. This automation may be well be a crawling phenomenon, but speeding up with widespread AI commercialization. We believe that we are very close to the point where people in society at large will join the dots between these phenomena. Our responsibility will be to steer the public uproar towards a real dialogue, to directions both truthful and beneficial, focusing on a shared civilizational vision coupled with pragmatic solutions. As discussed earlier, it is not the first time that humans have faced a challenge from machines, and there are many positive possibilities as well.

As a consequence of the magnitude of the coming shift, this era should be one of deep reflection – but that introspective process by no means equals idle chatter for years. There is no shortage of cheerleaders who praise only (or mostly) the massive productivity, efficiency, and profitability benefits of the AI Shift – but it is our intellectual and moral responsibility not to turn into pure hype-mongers. To put a first stake in the ground, it is our sincere belief that if this rise and proliferation stays uncontrolled or laissez-faire, it will exponentially strengthen negative trends we see at present and primarily empower actors with dubious interests. AI's beneficial outcomes are a *possibility* that we need to work for; some of its dangers are already a *certainty* that we have to manage and protect ourselves against.

The old World Order is crumbling, the new one has not yet been born, and labor seems to be very painful this time. Most descriptors of our age are "post-" this and that, but current AI is certainly "pre-", both in its own technological maturity and in its context, with its broader social environment being in flux. Especially in today's socio-psychological climate, the rebalancing act that we'll have to deliver as a civilization, on all fronts, is tremendous – and very much behind schedule. There is a huge magnitude of AI-related information available, but ungraspable to the public. There are big gaps in understanding even at the top echelons of power. We can see special interest-driven interpretations and omissions. In addition, the prevalent media practice continues to frame public AI discourse in a sensationalist way. All these factors acting in concert do not seem to help at all.

Another dismissive attitude and statement pattern we have experienced much too often is that AI is "just another wave of technology-driven transformation," that is, if "we as a society" figured out the combustion engine, oil, or data, we will be fine. (Actually, did we figure them all out? How do we measure figuring something out? By not going extinct?) Our work needs to demonstrate how wildly different this transformation will be, compared with all others that went before.

How is AI different? Our intention should be now focused on making sense of the distinguishing specifics of AI technology. Let us name a few key principles with the intent of busting some myths – all in the hope that the reader will find them helpful.

AI Power Principles

Principle of the Core

"So is AI like Blockchain or Virtual Reality?" This is a question I often get from non-technologists. Well, one good answer may be: "only from the hype perspective." A better one would be: "AI will merge with them all as the brain behind".

AI is the first general-purpose exponential technology: it is not one line of exponential technologies, *but the very core of technological disruption, and the primary driving force of the 21st century*. Rather than being one new technology to play with, AI is the transversal brain behind it all. It will blend with, and power all lines of technology, cut across industries, sectors, and all life phenomena in the foreseeable future. From the tech perspective, AI will be fused with and power fellow exponential technologies like AR/VR, autonomous vehicles or robotics. Emerging quantum and edge computing will increase the robustness and impact of AI technologies, most obviously by unbinding processing power and speed. Similarly, this approaching milestone will radically boost the path from "narrow-minded" general-purpose technology to general intelligence, combining knowledge across multiple disciplines into one single AI system.

But AI is even more than that. It goes beyond purely technology; AI mastery will become the primary source of power in our century. We believe that AI will be the core defining factor of the future economy, politics, culture, society – the big question is *what kind of AI* will get that overwhelming role and who will be in charge of it, which takes us to...

Principle of the Imago

Ultimately, Artificial Intelligence is nothing short of chasing a new Creation: the birth of non-human intelligence – and as the boldest would argue-, new "life." A statement of this magnitude may sound like it belongs to the realms of sci-fi, or theology, but it is very much real if we carefully delineate the controversial present from reasonable projections.

We know that we are still in the era of Narrow AI, but that is just an early phase. What is considered unanimously in AI research expert circles to be the next major watershed in AI development, Artificial General Intelligence (AGI) – meaning human-like cognitive capacity – will be created in the image (Imago) of its Creator.

Creator/Constructor are terms we will use interchangeably and broadly on purpose here, defining the scientist, developer, sponsor, business/political leader, organizational logic etc. who sets out to develop Artificial General Intelligence (AGI) according to their own morals, mindsets, interests, and priorities. AGI will project its Creator's power into the world and therefore will proliferate according to the Creator's objectives, interests, and ethical stance. Speaking today, even in its current Narrow AI incarnations that already surround us, the Imago Principle can be established. It may be less perceptibly so, but even in its present maturity, there are clear red flags that provide the solid foundation for many of our dilemmas in the AI ethics domain. Again, the narratives we are used to involve superheroes and supervillains and that may seem quite fantastic, but we are talking now about a technology that will follow its Creator's plan.

To sum it up; whatever goals an AI will be given, whatever will be programmed as objectives, optimizations and rewards by the Creator, will be programmed in his or her image. The version we get as consumers and citizens will be according to that Imago. Evolvingly autonomous AI will fill in the blanks in very creative – or at least, unexpected – ways.

Right now – on the path towards AGI – we can safely say that we should mostly direct our concerns at the Constructor's objectives and deficiencies *together* with the limitations of current mainstream AI technologies. This we will consciously do, rather focusing on the most intellectually titillating phantasms of "AI breaking loose." Before AI reaches super-intelligence level (full autonomy and therefore liberation) – if ever – AI technologies should be treated as smart, supercharged instruments expanding the AI Creator's influence. This especially shines through the current inability of AI agents to correct any design flaws and gaps that their masters have made - or omitted.

Principle of Unfair Dominance

The Law of Accelerating Returns was established by Ray Kurzweil back in the '90s and became the cornerstone of the Singularity ideology/pseudo-religion. In its original form, it establishes exponentiality as both the rate of technological progress as well as its fruit – be it productivity, impact, or power. Understandably, it became the fuel of the most recent wave of digital economic rally. The original statement is not without merit, since it admits that exponential acceleration creates a rupture in the fabric of human history. However, his final conclusion is that the Singularity's impact (e.g. the merger of biological and nonbiological intelligence, immortal software-based humans, and ultra-high levels of intelligence continuously expanding) is ultimately just continuous forced upgrading of ourselves by any means possible. Our (in)ability to catch up with the machines determines a path of accelerating power that leads us (or whoever survives) to the green pastures of Singularity. What we try to establish here is that the path is not all that rosy as this original vision – nor is it inevitable.

Our *Unfair Dominance Principle* means that the rewards of AI power are exponential. Out of the many competitors chasing the AI El Dorado, the one who reaches the big milestones will win big and leave contenders in the dust. We also argue that even if that dominance is only established in a Narrow AI field, the winner is likely to occupy that space and subsequently dominate it.

AI breakthroughs will not follow the historic linear pattern of tech evolution, not in their speed, power, or impact. It will also go beyond what is currently viewed as "exponential," especially our perceptions of exponential technology. AI evolves in leaps and bounds. In some Narrow AI applications (like cancer diagnostics or reviewing non-disclosure agreements), even in our Narrow AI Age, an AI can start from scratch, and only needs hours or days to reach and surpass human capabilities in a very narrowly defined task or domain.

When it comes to "instantaneous" AI's impact on the human brain and society at large, the decoupling of technological and mental/biological evolution will prove to be extremely dangerous in the absence of proper safeguards. We are already witnessing this in certain narrow fields where machines have bested the best of humans: chess, Go, social media impersonation, deepfakes, etc.... We can be outwitted. This disruption/disconnect/decoupling paradigm shift of humans left in the dust may expand continuously, ad infinitum. The concentration and expansion of power in the hands of AI Masters will follow suit.

The barriers of entry into the AI game are very high now and potentially will get higher; the resources needed to establish oneself as a *leader* even in one subset of AI technologies are only attainable for a very few. This rule applies even more to the AGI chase. So, it is not just speed: what actually became exponential is the cost of access to giant datasets, computing power, top talent, and gigantic live testing environments as well as underwriting the cost of research. If we are talking about AI computing power only, the cost of leadership is estimated to have soared 300,000 times since 2013. There is a fierce war for AI talent, and astronomical salaries are not unheard of.

Much of this race is zero-sum in the sense that whoever carves out dominance in a certain AI field or applied use arena (especially in the transversal effort, chasing AGI) will be hard to surpass and the capabilities gap may prove to be unbridgeable for new entrants, restricting access and keeping them down.

Others call this the 1% Rule. In the AI paradigm, maintaining even a 1% advantage in a given subgenre will yield exponential returns to the frontrunner. This rule seems to apply well to any kind of adversarial framework where 24/7 nonstop activity will be the new norm (see also Endless Siege principle below): cybersecurity, warfare, decision-support and forecasting systems, maybe other systems as well. If you are slightly better and can stay better, your rewards are not just slightly bigger but exponential.

Lacking broad competition, policy rebukes, or user rebellion, unscrupulous actors will have an incentive to proceed without major caution. This ”eyes on the prize” logic disturbingly permeates the unfolding AI race currently taking place in the global Wild West/East.

Principle of the Endless Siege

If you consider yourself a self-perceptive person, you may already have a certain level of unease concerning the level of digital intrusion in your life. Well, the bad news for you is that it will get exponentially worse in an AI-driven paradigm. All characteristics of digital influence – depth, precision, sophistication, persistence – will be augmented by several magnitudes, optimized for maximum efficiency.

Imagine that systems with seemingly endless resources can monitor you 24/7, seeking your vulnerabilities to extract data from you, to sell you something – be it a commercial, political, or religious product. Tireless systems will continuously and subtly urge you to do something, or to refrain from doing something. If we think broadly about it, this is the very definition of influence or power.

In one near-term dystopian scenario – that we as a society are already experiencing to a certain extent – Orwell's Big Brother becomes the Big Salesman. Habits can be changed and built with well-timed nudges – and who wouldn't want lose bad habits and build good ones? From habits, the next frontier for marketers or their AI to influence and overtake will be instincts, emotions, our needs and desires as biological creatures... and ultimately, the mind and the will. Every single reaction users produce will optimize the algorithm further – not necessarily to your benefit. Once it finds a point of entry, it gets under your skin. Your room for maneuver in options being considered, and decision-making, will be reduced: your resistance to manipulation will be lower and lower – until it is gone.

What if we go beyond the impact on the individual and talk about the systemic: consider macro-level impact such as security. Especially in the context of cyberwars, the Endless Siege will be very much literal by means of AI technologies. According to cyber defense experts, the siege is already on in our current global security climate, which we sometimes call the "LukeWar(m)." Ours is an age defined by continuous "shadow wars," without either enduring peace or all-encompassing acts of war. As "unconventional" tactics become the new norm and AI revamps the military arsenal of projecting power, our attention and imagination are often drawn to killer drones and their ilk: this is understandable. Nevertheless, even if it is much harder to grasp conceptually for the human psyche, we think that zero-casualty, zero-mushroom-cloud, endless cyberwar is no less dangerous for society.

Principle of Horror Vacui

Horror Vacui is a Latin expression derived from Aristotle's "nature abhors a vacuum," literally translated as "fear of empty space" and historically mostly used in art.

Currently our minds perceive contemporary AI solutions as being only narrowly focused on certain tasks, industries and applications – hence the term Narrow AI. Since the aim of AI development is defined as the full replication of human cognitive capacity by machine intelligence, this target of "general purpose intelligence" is at the very root of related efforts. Based on current trajectories, it is highly plausible that general-purpose AI agents will be vastly commercialized and deployed long before the theoretical goal of full human cognition replication will be achieved – if it ever is. Regardless of how long the period of these "AGI mongrels" will take, we have to be prepared for full proliferation before perfection. The goals of AI creators to market and scale AI agents meets the innate characteristics of AI technology for expansion and optimization. This will destine AI technologies to operate across the board and fill all the nooks and crannies of our digitized lives.

As mentioned in earlier chapters, we have reached a point when some voice-operated home assistant devices are available – sometimes offered for free. They are to be stationed in the rooms of your house, collecting voice and all other kinds of data. As humanity is slowly waking up to it, this data is a lot more valuable than the hardware. The current competences of these AI assistants may seem narrow, buggy, and meager. However, through continuous usage – and often unknowingly – we nurture them to become "general-purpose": we personalize them with our data, helping them to broaden themselves; they get smarter with imperceptible upgrades and deepen their understanding of our needs, all simultaneously.

To be clear, we would like to argue that AI's unique characteristics of self-improvement, self-replication, and autonomous evolution ensure a continuous, real-time expansion of personalized AI technologies in our lives. Their presence will soon feel indispensable as our anchors and guides that get us through the day.

Chasing AI leadership (especially that of state actors and to a somewhat lesser extent, tech leaders) is often compared to a nuclear arms race. That may be appropriate regarding the operating mindsets of these direct AI competitors. But on a higher level, regarding the breadth, near-term ubiquity and impact of AI-infusion on all of our lives, we believe it is much closer to electrification regarding its strategic importance, omnipresence and coming indispensability. How did you react to your last power outage?

On the Origin of New Species

The current AI narrative in the public mind is beset by the problems of obscurity, opacity, and obfuscation. This results in an inability – or outright refusal – to prepare for the future. Society needs to make delineations and definitions to address the currently visible path of AI maturation, as well as the growing uncertainties on that forecast path.

Disjointed narratives try to cast our attention either in the far too distant future, or on the often tragi-comical shortcomings of the present state of AI. News media like stories about trouble, especially in the era of chasing clicks. Media have promoted stories on how easy is to hack autonomous vehicles' object recognition by using tape on road signs, to the more sinister discovery that a faulty facial recognition system mis-identified African-American Congress members as criminals. Make no mistake: talking about AI shortcomings and failures is the responsibility of the media. However, what matters is what the reader will ultimately think about the personal relevance of the issue. These extreme media narratives may imply some sort of irrelevance (too far in the future to matter) or untimeliness (not yet fit for purpose) for the AI societal discourse. In other words, according to the rules of engagement of the current media paradigm, trending stories are either about pathetic fledgling AI mistakes, or are stories of mass enslavement by a certain Supercomputer of a grim future that we feel no connection with. These extremes may hamper our societal capability to get up to speed.

In our sense-making and rebalancing journey, we must make an attempt to put some stakes in the ground regarding definitions and timelines. Please note that these definitions are not sacred, but it is hoped that they will help our common understanding for the remainder of the chapter.

For an unusual type of framing, at the end of the second decade of the 21st century AD, we are in the second 60 years of AI development. Using 60 years to determine a cycle sounds arbitrary but practiced traditionally by the people of Mongolia[4]. Mongolians measure history in 60-year cycles, called a *jaran*. The fact that it is the "2nd AI Jaran" needs to be emphasized for two reasons.

One is that this long-term view is juxtaposed with the ephemeral worldview dominating our digital society – of its shapers and consumers alike. Our current paradigm of experiencing time is determined by the "stream" phenomenon, where everything is in constant flow: in this flow, we are conditioned to constantly crave "what's next," what is going to happen in the next few seconds. Philosophically, while it may be alluring to equate this to the "being in the now" principle of mindfulness, in practice, it is nothing but mindless and frantic – so becomes the exact opposite. In this directionless paradigm, cycles of hype and bust are also approximating real time.

To reach an adequate vantage point to observe the vastness of AI's potential and impact, it is worth citing the various timelines clashing in the AI context: the long jaran-like historical cycles; the decades of digital transformation (think terms like Fourth Industrial Revolution and beyond); the cycles of "AI Winters" and hypes; the exponentiality of frontier technologies spearheaded by AI; and last, the endlessness of its transformational promise, once it matures.

Our number one caveat is that if these timelines diverge too much, i.e. exponential AI power meets the unbridled Data Economy, industrial organization of labor, medieval institutions and troglodyte instincts, the future seems very dystopian. Society must keep pace with the developments around it and infiltrating it. An across-the-board conscious upgrade with concerted governance is the only way all of us can make this work for humanity.

In the world of fiction, the sci-fi genre kept AI on the agenda. However, in the waking world, the general public was mostly unaware of the fact that AI development was going on for many decades, in AI surges and busts – the latter are called "AI Winters." (Our tongue-in-cheek definition of an AI Winter is "When we punish ourselves for overinflated expectations.") For most, AI became a reality when the media started covering it extensively in an oversimplified way, but that started years after AI-powered products and services had already hit the mainstream consumer, enterprise, and government markets.

As we have described, for the last decade or so, we have been in the **Mainstream Weak/Narrow AI Age**. But as AI research veterans would say, we have been here since the 1950s, so what has changed? After quite a long AI Winter of discontent, the current AI Shift was caused by a first-time celestial alignment: the arrival of cheap and massive amounts of data together with both hardware capacity and algorithmic breakthroughs. These factors together exponentially increased the robust computing power that AI needs, and that we were historically short of.

While humankind at large is barely waking up to it, Narrow AI is now mainstream, commercialized, widely used across industries/sectors and therefore influencing our lives. Most personal experiences and memories of consumers interacting with AI from this Weak AI Age will be through online consumer chatbots, language translators, or AI voice assistants of an ever-growing scope and acumen. Those with a penchant for observing undercurrents will start registering (and hopefully, analyzing) the effects of product and content recommendation systems, social media manipulation and disinformation, stock trading robots, decision support solutions – and, perhaps most visibly in the media – autonomous vehicles, on their lives, as we ramp up for the next stage. Largely unnoticed by the public, mass facial recognition systems are sold to law enforcement and defense agencies and have been operating for years now. On a macro level, some emerging powers are piloting the first modern surveillance state and run the whole country on an AI "operating system."

It needs to be stated, though, that the current dominant technologies are machine learning based (e.g. pattern recognition, computer vision, neural networks, natural language processing etc.) and are still growing up. Machine Learning technologies are fragile and have severe limitations related to data quantity and quality; stochastic to deterministic models; ethics and explicability. They are also very resource-intensive (with a very visible and expanding environmental footprint). These are just the most obvious limitations. Research frontiers are now exploring different architectural models, as well as ways to make the engine that powers AI leaner and nimbler – with results still to be seen.

In their candid moments, AI researchers and developers acknowledge that oftentimes, most currently commercialized paths of Narrow AI do not seem to lead towards replicating human cognition and will be considered dead ends from that science-driven aspect. But what matters here is that they produce significant economic value already: despite these obvious shortcomings, Weak AI is creeping forward and proliferating unstoppably in both the commercial and policy realms. The big question is intent and direction. Besides already generating significant revenue and projecting power, will these current incarnations of AI lead towards a humanized AGI, and will they ultimately benefit humanity, or do we need major course corrections and a refocus on other AI approaches?

New experimental structures and architectures, especially those built with adaptive, evolutionary principles, to learn organically and unsupervised, seem to be the most promising directions in the pursuit of human-equivalent cognitive machines. The danger is that the road to there will be bumpy and we will experience a phase of "Dr. Moreau's AI" in deployment. This may well happen due to a perfect storm of several factors: short-termism of quarterly earnings, the pervasive competitive race mentality, and the fragility of machine learning, amongst others.

Even if we end up on the more prudent side – also because of our innate human tendency to anthropomorphize – it is best to treat emerging AI technologies as children, with their unique developmental needs. Google's DeepMind's IQ was estimated at 47 in 2017, with a progression objective of 20 points over 2 years. So it is now, in 2020 a little older and smarter than a six-year-old child (IQ=55).

Servants and Monsters

Once those children start maturing, they become "enfants terribles," aka **Strong Hybrid AI or pre-AGI (AI 1.5)** In other words, AI technologies will grow powerful, resource-gluttonous, pervasive, convincing, and in some cases, intimidating to humans – but stay flawed and incomplete.

One of the iconic examples of this in-between phase – that may allow some to argue we have already reached that stage of hybrid systems – is IBM's Project Debater. It is an AI *debatteur*, which can deliver surprisingly good argumentation purely based on compiling and filtering vast amounts of data – but without any cognition. Some AI experts argue that Google's AI, diffused and distributed across a wide variety of applications, is also at the hybrid pre-AGI stage. Google's AI engine is already capable of writing software, and that trait is borderline AGI territory when it comes to autonomy and proliferation.

In sum, we are now on the verge of the age when the first major global blunders can be caused by half-baked AIs with too much autonomy (think AI equivalents of Frankenstein's monster or Caliban) and lead to creeping negative transformations – or outright short-term catastrophes. For 2020, major analysts see a 70%+ chance of at least three major global PR disasters caused by AI technology that is already deployed.

For one example, a key frontier of AI development is enabling human-like conversation, measured by a Sensibleness and Specificity Average (SSA) score. Google's most recent Meena chatbot claims conversational AI leadership because of its 79% SSA score, 25% higher than any other conversational AI models. Google says that the average human SSA is 86%, so Meena gets excitingly close to that. This technology may come to a home device or phone near you very soon.

Every week there is some breaking news story of a major Tech company achieving an AI augmentation or breakthrough in a subdomain that is dubbed human-equivalent, or even surpassing human capacity in a more cross-cutting domain and leading further on the path to AGI. This leads to dilemmas like the following: if a human professional role has four major work tasks that an AI can fully replace, will anybody care that it is not proper AGI from a scientific point of view? However, in the eyes of the employer, it definitely constitutes full replacement for that specific role.

This is the age of hype and hysteria. In this ruthless AI race, AI leaders think, keeping up appearances is of paramount importance. On the one hand, companies worldwide employ millions of low-skilled human workers to bridge the current gap of AI's shortcomings – or even, cleaning up after AI's outright failures: this is affectionately called pseudo-AI. For example, largely unnoticed, there are armies of contractors helping out social media AI detecting fakes or hate speech or doing generic data labelling – the equivalent of digital menial labor.

On the other hand, massive AI solutions can be deployed, rather unnoticed and unregulated, that should have stayed under wraps at the lab for a long while. "Build the software until it is 80% ready, then go to market," – this has been a popular quip in the Tech C-suite. As opposed to "old world software," where users could report bugs that could be patched, the reality with AI is that there is no room for that kind of risk-taking or leniency anymore.

As some leading AI safety researchers put it, if it is 99% ready, it is 100% dangerous and absolutely not ready. We are nearing a crossroads when emerging AI will produce servants, companions, or monsters. Our joint responsibility is both tremendous and urgent in setting both actionable directions for AI safety measures as well as ongoing AI governance checks and balances without further delay.

Python and the Grail

As we hinted at above, the Holy Grail of AI research (and that of the intensifying, no-holds-barred race) is **AGI – Artificial General Intelligence (AI 2.0)**. In its magnitude but not in its speed, getting there from the current Narrow/Weak is an evolutionary jump similar to the one from the circus monkey to the *Homo Sapiens*. AGI (also known as **Strong AI**) is defined as the capability to fully mimic generalized human cognitive abilities in a machine setting, so that, even when faced with an unfamiliar task, the AI could find a solution.

In theory, an AGI system could perform any task that a human is capable of, and therefore potentially replace – or fully support – a human at a wide variety of tasks, indistinguishably from the human agent. The definitional nuances accumulate when it comes to choosing between emulating/replicating/replacing/mimicking as the verb describing the developmental objective.

It needs to be stressed that most AGI definitions only focus on the utilitarian approach of "task replaceability" or level/amount of intelligence displayed; they care less about whether tricky terminologies such as "mind," "common sense" or "consciousness" have been at play. Current and near-future AI systems based on deep learning demonstrate a brittleness – the AI is very good at registering patterns unseen by the human mind, but is poor in adaptation to changing data environments. Acquiring generalized knowledge (that is far more than a database) and the capability to establish analogies will be a crucial milestone on the way to full AGI. Our human common sense sometimes overrides even specialist knowledge – current AI is incapable of that. Building AI systems based on "common-sense reasoning," which replicate the uniquely human ability to amass contextual knowledge based on experience, is a very resource-intense and complex activity – but will be the only truthful way to AGI.

AI optimist experts see their own definition of AGI emerge by 2030, broader consensus among the more rigorous is around 2050–2060 – a minority says that it will never happen. But remember, AI scientists are usually a lot more rigorous in their definitions than salespeople, startups, and journalists.

A very distant possibility on the horizon – but also the one capturing the imagination of Hollywood the most – is Artificial Superintelligence – ASI (3.0). As the ultimate, controversial frontier discussed widely by experts – and Hollywood – ASI is when AGI evolves beyond its intended original raison d'être set by the Creator/Constructor, escapes from the box, starts self-replicating, mutating, shaking off its yoke – and becomes fully autonomous. Defined as the point when AI's capabilities surpass that of humans (and that means all humans), the most disconcerting trait of ASI is that while we may have a good guess on the time needed to get to AGI from here, the jump from AGI to ASI can potentially be measured in minutes. In a similar vein, it needs to be stated unequivocally that the "ability gap" between humans and ASI follows a similar pattern: it gets exponential in a matter of minutes due to self-learning, assimilation, and self-replication, only hindered by access and computing power. For that very reason, when preparing society for AGI, we cannot decouple ASI from it and we have to approach them as a single package – even if our intent to stay out of sci-fi-esque speculation remains unaltered. This is a point often captured by human imagination – both scientific and fiction – so we kindly refer the reader to these works.

The Moment Before the Tempest

> "Artificial intelligence is the future for all humankind. It comes with colossal opportunities, but also threats that are difficult to predict. Whoever becomes the leader in this sphere will become the ruler of the world."
>
> *Vladimir Putin*

As we attempted to demonstrate above, as responsible thinkers, we can't be mere cheerleaders saying, "let us build AGI and we'll figure it out on the way." At this crucial crossroads in history, we need a firm vision on *what kind of AI* we want. For many, this seems like a purely techno-economical set of choices, but we believe that it is more of a moral-philosophical one.

As of early 2020, Weak/Narrow Al is already here, heavily commercialized and deeply influencing our lives, households, relationships, transactions, and work – without most people noticing either its presence or its power. Instead of musing about the boons and banes of Artificial Superintelligence, there is already a treasure trove of contemporary or near-term phenomena worth our analysis.

So how much of it is out there? While the exact definition is not easy, Al is foreseen to become a $60 billion industry by 2025. By the end of 2020, there will be close to two million *new* AI-powered robots that be will joining the industrial workforce as new "employees," tracking the last two years only. AI is also domesticated: by 2022, 55% of global households worldwide are expected to have an AI voice assistant. And while we are on the go, 80% of smartphones are projected to be equipped with on-device AI capabilities by 2022. The channels of deployment are therefore set up and ready, with constantly evolving artificial intelligence solutions establishing two-way data flow and interaction.

Only the robustness of the AI brain behind will change – quite fast and unperceived by most. Consumer solutions that help us seamlessly play music in our bathrooms, come up with cooking recipes instantly, or help us navigate from here to there are widely accepted, used, and deemed overall positive and even indispensable by consumers. Similarly, automation in the workplace so far is perceived mostly positively both by employers and employees, but with growing fears of being out-automated, or dehumanized.

Data is the foundation and fuel of an algorithmic paradigm. But given the low level of data literacy in the general public, combined with the lack of appropriate regulatory safeguards, the complex societal challenges of AI will have to be tackled while being subjected to an already uncertain and fragile data paradigm. The voices defining our data as our very *digital imprint* – and therefore protected at the same level as individual rights are protected – are growing stronger but not yet matching the pace of evolution, deployment and data-hunger of AI systems with appropriate regulatory and governance frameworks.

Without the right level of data literacy and understanding in place, it is easy to overlook the inflection point here – even if we stay constantly vigilant about the hype factor behind AI. Due to data and AI being indivisible and inseparable, we need to stress the following: we may well end up being stuck on the current exponential trajectory and end up at an undesired and unexpected point, at the mercy of those who were preaching that all AI “bugs” will be fixed as necessary, but they can not “stifle innovation”.

The proliferation of AI algorithms, robotic applications, and consumer-facing delivery channels are a done deal. Commercialization of whatever AI technology is deemed market-ready will be unstoppable, unless major catastrophes and backlashes occur. New releases will reach us imperceptibly through our devices already in position. Our reliance and dependence on – and sometimes, infatuation with – technology is becoming a smooth, somnambulant stream. Simultaneously and as a first in history, AI will take care of optimizing the selling of itself – and of everything else that needs to be sold, in parallel and continuously. “The customer is always right” was the mantra of the Industrial Age, but with AI-driven advertising brought to perfection and getting under our skin, do we customers still have a choice – or do the rules need to be rewritten?

AI tech itself could best serve the purpose of forecasting the impact of this transformation, but it is not currently used as such: who would take this on? Lacking that kind of pre-emptive thinking and joint effort, orchestrating a human-centric vision and ethical consequences will be an afterthought. The pace is dictated by industry and bellicose nations, and there is a gap that is currently unbridged with policy, when it comes to mutual understanding, trust, and joint, actionable vision building. This setup is a ticking bomb for all parties concerned and needs to change rapidly.

The big question is: what kind of AI gets control over our lives, how can this control be managed and counterbalanced – and ultimately, who is in charge?

CHAPTER 4

THE GLOBAL RACE FOR AI DOMINANCE

Scrutinizing the Imago Principle further, we can state with certainty that a few key players will dominate at least the next phase of the evolution of AI. The reason for this lack of diversity is primarily derived from the critical mass of prerequisites that set the barriers to leadership so high: robust infrastructure, access to/ownership of primary datasets, and the ability to attract elusive and scarce talent being the most important ones.

When it comes to **state actors**, there has been a visible explosion since 2015/2016 both regarding the strategic priorities defined and government investment in AI. Following a logic similar to the nuclear arms race that started in the mid-20th century, but at an even more accelerated pace, AI has lately been grabbing the attention of a vast set of mid-level powers, while the big powers further solidify their grip on AI dominance. There are 60+ states who have published national AI strategies as of today. What is even more interesting is that even more – an estimated 80 states – are working on autonomous weapon systems, the most controversial edge of military AI applications. According to disinformation watchdogs, about the same number of states are involved in state-backed, AI-powered disinformation operations. There are numerous questions associated with the risks of surveillance/facial recognition AI systems, which tend to be the least publicized but most prevalent procurements of the last few years. The race is on, and national security use cases are taking precedence over the much slower overarching societal AI strategy, which will have a lot more creeping transformations to deal with.

For some of the biggest AI powers, their AI focus and investment is heavily influenced by what others invest in. In line with our Unfair Dominance principle, this is primarily driven by the objective of reaching supremacy in their projection of defense AI applications. Secondarily, we see big efforts in AI surveillance stacks (facial recognition etc.) and in forecasting systems. To some extent, this is understandable since there are clear and present dangers upon which one needs to act decisively, but there are layers of creeping societal transformation that need a different mindset and pre-emptive thinking for an appropriate strategy. Examples of the latter include the implications of AI and automation on the world of work, the impacts of AI on democracy and the public dialogue or the invasiveness of AI-powered advertising on citizens and consumers.

In sum, we believe that worldwide more emphasis should be given to the government AI stack that primarily affects citizen well-being and the efficiency of government services as parallel objectives.

Historically, the ***US*** has led the world in all aspects of AI research, security use, and commercialization since its very inception. For a long time, it has done so without any serious challengers to both its model and unparalleled resources. To name just one aspect, historically, US investment into Al has amounted to around two thirds of the global cake. In recent years, the visible trends in how the US perceived its own AI interests were mostly following the double imperative of:

- Keeping the supreme edge in the defense aspects of AI;
- Guaranteeing algorithmic economy primacy for the US-originated global tech companies that are the unparalleled leaders and investors in AI.

We need to dig a little deeper here for a more nuanced understanding. The task at hand now for the US is a complex one. Some related threats and difficulties are domestic, for example:

- The uphill battle of regulatory dealing with the tech monopolies, and devising the federal data protection policies that are a prerequisite of any actionable AI societal strategy;

- Protecting US institutions and public dialogue from the devastating fragmentation caused by AI-powered disinformation/manipulation campaigns;

- Keeping a tight balance between national security interests and having the leading edge in AI surveillance technologies, but sticking to American values, freedoms, and rights in doing so, thereby avoiding the more questionable path of global competitors who try to dictate the pace;

- Protecting US AI technologies from competitors: for example, the AI aspects of the escalating trade war (i.e. ban on certain AI software exports) are also very visible and stand a good chance of getting stricter.

There is a global ebbing of public trust in institutions that has been clearly measured: US democracy and institutions are no exception from this global phenomenon. Fortunately, it is in the resilient essence of democracies to shift and reform themselves in times of trouble. It is our belief that the fundamental values and citizens' rights on the US must be kept robust and protected as the country adapts to these new technologies and will inevitably be exposed to many temptations.

Until very recently, we have mostly seen tactical statements coming out of the federal government that urge other powers not to overregulate AI (i.e. "stifle innovation"). Lately, there is a welcome, emerging focus on the ethical aspects of AI that government agencies are expected to observe in their procurements and regulatory activities, as instructed by the White House – we hope this was the first of many statements that will go deeper and deeper into the ethical and responsible usage and regulation of AI.

The US is the number one power in AI, as well as the strongest democratic state and the leader of the Algorithmic Economy. We believe that there is a set of responsibilities that come with that. As a turnaround in recently emerging anti-multilateral tendencies, the US would be advised to pull its weight in contributing to the emerging AI ethics, norms, and institutions that transcend borders. The AI future needs moral as well as technology leadership. Using its traditional focus on innovation unhampered, the US should take a leadership position on AI innovation, both by mobilizing the private sector innovation ecosystem with the right incentives and by refreshing the state-backed innovation models that built Silicon Valley in the first place. Building novel frameworks that do not jeopardize its AI supremacy, but also benefit allies should be key. This way, the US could be viewed and respected as an enabler and model-builder for the world. Such a move could meet the EU's interests as well and somewhat revive the transatlantic alliance, especially in light of the partnerships-first stance the European Commission took in its AI White Paper.

The international community expects US leadership and contribution to a global AI governance vision and model. To support such an endeavor, the global multilateral institutional framework needs conscious steps from the US that reform and do not dismantle it. US national security largely depends on building and projecting a model of AI-conscious digital democracy to the world and rallying temporarily disoriented allies behind it. Lately, promising signs have been emerging for the US to put its weight behind the Organization for Economic Cooperation and Development/OECD's AI Policy efforts.

In parallel, this effort would also be essential to reverse a dangerous emerging trend that pulls uneasy mid-sized powers into the sphere of influence of the US's biggest competitors. The closest rivals operate with a very concentrated, "all-in on AI" strategy and with a preference for asking far fewer questions regarding AI tech misuse and the ethical consequences for their customers. For an apt example, the unfolding reluctance of even the closest political and military allies of the US to say no to Huawei is perceptible, even though the US made this wish clear in no uncertain terms. The world needs a US-led AI strategy and global framework that includes allies as partners – and this does not necessarily have to go against America First in principle.

Compared with a large number of allies and competitors, the US was late coming up with a comprehensive, public AI national strategy, first in the shape of an executive order in the spring of 2019, followed by a more strategic document in June. The first government-targeted set of AI ethical directives came out from the White House in January 2020, in the form of a cross-agency memo, without too much fanfare. The US is a late bloomer when it comes to AI public strategies, and federal policy directives are still in their infancy: reasons may include a purely market-driven philosophy or even that of the administration seeking to define its role. Critics say that in its present shape – especially given the global weight of the US in research, tech, military, etc. – what is visible today in the realm of responsibility will need a lot more work from a leader.

In line with the traditional market-driven American worldview, as long as US companies dominate the AI race technologically, the federal government may equate that economic lead with US strategic leadership. Maybe the same type of thinking from the administration also applies to what other American stakeholders are doing, such as MIT's billion-dollar investment into an AI "super-campus." The only area where direct government investment is clearly articulated as such is in the funds allocated to DARPA, to be deployed for national security purposes. However, DARPA is also responsible for research that goes a long way beyond purely military objectives. For two quick examples: DARPA has its Machine Common Sense program (we touched upon that aspect earlier): in short, they are set to eliminate the most significant barrier between Narrow AI and AGI by building experiential computational models coupled with a system that learns to research the web for a "fluid knowledge repository." DARPA is also expected to achieve breakthroughs in Explainable AI (XAI), one of the toughest areas and regulatory barriers of AI. (We will discuss XAI later in the book.)

Recent developments show some friction, though, in the domain of AI defense projects with the private sector.[5] Some key government AI projects have been canceled, and some others are on the brink of cancellation due to an "ethical revolt" of many Big Tech employees. These workers chose to say no to both their own management and to the government (both federal and foreign) as a customer.

Is this the start of a trend? It is hard to tell at this point whether this new phenomenon will subside due to Silicon Valley CEOs quelling the revolts, the government putting a more conscious effort on the ethical to appease critics, or through other, more willing companies stepping in to fill the gap. But it is also an intriguing question whether this "talent rebellion" can become a key trend of our times. Lacking a unifying external threat that is perceived similarly by all, it seems likely that the growing cultural divide between Silicon Valley tech workers and the federal administration will further fuel this friction, at least in the short run.

Such problems are unknown to ***China***. Setting a new course in 2015, the rising superpower has moved beyond Phase 1 of its grand strategy to become the dominant AI global leader by 2030, according to a centrally orchestrated national masterplan. It should come as no surprise, then, that in 2018, year-on-year Chinese investment into AI has exceeded that of the US for the first time ever. To illustrate Chinese AI focus with an oversimplified historical reference, China's vision could be summed up as a close relative to Lenin's 1920 strategy for the electrification of the Soviet Union: Chinese state capitalism is the Party plus artificial intelligence. In line with the specific rules dictated by Al development, the blurred lines between state and industry, the size and depth of its primary datasets, as well as the capacity to focus due to a centrally planned digital economy seem to yield advantages for China. These factors did not play in its favor with previous waves of technological change, but because of the nature of the AI power principles that we laid out earlier, they do now.

When it comes to public opinion, the Chinese are the most AI-optimistic nation on the planet. According to a 2018 Ipsos study, 62% of the general public labeled themselves as AI-optimistic, almost double the world average (35%). China sees AI proficiency as the prime vehicle to a supersonic rise on the world stage and the population seems to share the Party's wishes to be recognized as No. 1 in AI, even if there may well be clear and present trade-offs in liberties they need to give up as citizens.

Instead of a "spray and pray" tactic that historically haunted both EU and US public investments, China chose to pick winners both when it came to startups, and handpicked homegrown Big Tech companies who were selected to occupy certain sub-segments of AI development and commercialization. The world's best funded startups in Al are Chinese, and a lot more Chinese AI flagships are part of the Top 100. Not purely by accident, China's world leading

AI startups SenseTime ($7.5 billion), Megvii ($4 bn), and Yitu ($2.4 bn), were blacklisted in the US. Chinese AI social media leader TikTok (valued at $75 bn and boasting more than half a billion users) is now banned within the US military, which cited national security concerns.

The vision dictated by the Party is built on pillars of equal importance: an AI-powered surveillance state (a term employed by critics) fueled by a social credit system, a hybrid cyber/AWS military force and commercial mega-startups aligned with (and in many cases, owned by) state-sponsored Chinese tech giants (e.g. Baidu, Tencent, Alibaba). Many experts say that Chinese AI strides already achieved in the plan will be difficult to match or surpass, even if much of their impact remains behind the veil as of now. Realizing the dangers associated with that rise, the US and several western powers are now stepping in, blocking mergers and acquisitions, and even banning the import of some flagship Chinese technologies outright as a very important strategic subset of the unfolding rivalry.

When it comes to soft projection of power, Chinese investment in Silicon Valley startups is already scrutinized by the US as a potential security threat. The web woven by China globally by investing in dramatic magnitudes into the infrastructure of developing countries enables it to occupy leadership positions left empty by the post-Cold War power vacuum. This well-established activity foreshadows a game of encircling by occupying Tier 2 commercial tech – including AI – markets as an emerging No. 1 vendor, as well as influencing AI government tech systems of these countries to be built under a Chinese umbrella.

The ***European Union*** as well as its member states have a historic opportunity to define a "European AI Way," when it comes to regulatory foresight and social vision of AI implementation. Under the

European flag, there is a lot of deep thought fueling various (concurrent) AI-focused alliances, committees, and projects launched inside and around the EU institutions for years now. Europe took its time to handle the intellectual complexity of the task at hand: that is, regulating the most fickle set of technologies ever while seeking paths to establish "EU AI Sovereignty." When setting out to regulate AI, Europe's social vision remains firm and its steps cautious. In the European thinking, betting on AI and tackling threats proactively are equal mandates and with the latter, it set out to protect society as a whole, as well as the individual. This double imperative is expressed in the will to build what the EU frames as AI Ecosystems of Excellence and Trust.

The intent to make AI socially viable and sustainable by design is a promise for Europe's digital future and its emerging model of "strategic culture." The new Commission that came to govern at the end of 2019 named AI regulation as a top two priority for its first 100 days, and an augmentation of EU funds going into AI is foreseen from 2020 – however, it will still be dwarfed by other major powers. Admitting its currently unbridgeable weaknesses in consumer AI platforms and datasets, Europe is planning to focus its basic research and industrial strengths and capabilities on exploring other directions of AI development (for example, bridging ML and symbolic AI approaches) and betting on edge computing, in an attempt to leapfrog rivals. All that is commendable, if aspirational, on the vision end. Similarly, the Commission's plans to open up high-value public sector datasets to create a €6 billion European Data Space to reduce its data exposure and vulnerability is a major step for increasing its AI prowess and maneuverability.

Europe's focus on a well thought-out AI regulatory framework is to be highly desired. The Commission sees the red flag, though: Europe has good research fundamentals, but keeps losing momentum and brains to other powers, especially the US, where top-notch AI researchers and startups flock in droves. Also, on the regulatory end, there are some cautious mentions of regulatory hurdles that are almost impossible to manage with the current policy toolkit, which has no agile/pre-emptive elements: namely, the inability to follow frequent software releases/updates, and, regarding AI, the very nature of learning technologies being elusive. Most importantly, emerging AI regulatory frameworks will be risk-based and in that aspect, the EU points out both high-risk sectors and high-risk use cases: for example, Europe names recruitment AI and surveillance AI as two high-risk areas, calling for risk-appropriate regulation. This approach – if adopted – will certainly aggravate clashes with AI industry leaders as well as with other AI leading states.

However, one can only wonder if the sheer lack of pace in implementation will not relegate Europe into a Tier 2 AI player. Europe historically struggles with what is known as the Extended Innovation Paradox: namely that immense public funds spent on innovation rarely find the path to mass commercialization. In this vein, it is equally problematic as a persistent issue that Europe is lacking a single homegrown tech giant: this will be further aggravated with the lack of an AI giant like the ones in the US or China. (The same can be said about the historical baggage of not being in control of the large EU-sourced commercial datasets that are fundamental to building up AI leadership).

If you are confined to scrutinizing technologies made outside of Europe and commercialized from elsewhere, and your weapon of choice is slow-moving regulation, the omni-stakeholder trust that needs to be built for the vision of a human-centric AI Age may fail to emerge. There is a version of a “cold” trench warfare between the EU and Silicon Valley, with a dire need for mutual trust to be built up. Therefore, Europe needs to broaden its perspectives on modes of intervention: there are some promising early signals for that in recent documents coming out of the Commission, especially regarding global partnerships and closer cooperation with industry.

If Europe wants to play a leading role in the AI Age, a series of steps will need to be taken, especially in the field of resource concentration. The EU needs a revamped and much more focused funding system, regarding both commercial and public interest AI technologies. Europe should put the right incentives in place to lure (back) top AI talent. The technological gap between that of leading EU companies and global pioneers cannot be closed purely incrementally. We stay hopeful that Europe will shake off a potential spiral of complacency, and act both wisely and decisively.

Setting a social AI vision at the EU level, coupled with forward-thinking policies will be essential, but execution will be the ultimate challenge. It is probable, though, that progress may be slowed by structural issues, such as the diverging interests of the member states who have each developed their national AI strategies in relative isolation. Calling EU big bets in AI investment will be made more difficult by the fact that, for example, the planned magnitude of the EU's short-term investment in AI is in reality very diffuse, spread over a large number of grant schemes. The whole EU scale of investment planned at the end of the previous budgetary cycle equals that of France's national AI investment alone. For a distant comparison but hinting at magnitudes in "bang for buck," it also equals the budget for Google's new campus in New York. The Commission's proclaimed separation of civilian and military AI – with the latter left completely untouched by their evolving AI policy – could also prove to be dangerous and increase Europe's defense fragility. A lack of a proper AI cooperation framework between the EU and NATO will further aggravate the vulnerabilities when it comes to AI cyber-defense aspects: some adversaries are already exploiting this disconnect ruthlessly.

Despite these enormous burdens, it needs to be further emphasized that the European AI way may become alluring in the longer run for talent worldwide. As of now, Europe seems to the only major conglomerate on Earth that is putting the ethical and the social transformation issues of AI front and center. Europe may be capable of projecting a forward-thinking, visionary model with the potential of steering Al in a beneficial way for the whole of society, based on strong, humanistic social values. We can only hope that it does not come too late – or in much too diluted a manner.

Russia's leader is often quoted as one political actor who deeply understands the significance of Al as the primary source of power in the 21st century. This alone should send warning signals to democratic policymakers following a more laissez-faire approach, for one reason or another.

Russia is no AI Superpower but needs an "honorable" mention. Its means are significantly more constrained when it comes to public-facing AI investment (estimated $600 million total), but its much fewer cards are being played for maximum impact. It published its national AI strategy at the end of 2019 as a very late bloomer, but not too many surprises there. You don't need to be an expert Kremlinologist to say that after all, that document is less important compared with the way Russia works in practice.

What is important, though, is the unspoken expert consensus that is routinely denied by Russia: its skill in using AI to manipulate the minds of people in foreign lands. The last few years have clearly demonstrated that concentration in a few narrow fields of Al yield a Russian capability of projecting hybrid power that far exceeds its economic prowess and global status. Simply put, Russia does what it knows best, and has mastered the art of Al-driven hybrid warfare (cyber/intelligence/infowar) in a few years. This does not mean dominance, but strong disruptive competence and deterrence. This specialization will continue to be a source of incidents of a growing magnitude that will serve as pilot projects for them – and hopefully, as wake-up calls for the international community. Chances are that smaller powers with resource constraints, but without citizen control, will be inspired by the Russian model. We have already mentioned 70 states with active disinformation efforts as of September 2019: these programs have doubled in a mere two years.

A bit more thinly veiled vis-à-vis the rhetoric of state players, the AI race between Big Tech corporations is no less fierce. The logic, the aims, and the means are different though.

The big companies, the ***Tech Titans*** possess everything to win the game, and they are making big strides in solidifying their AI ecosystems. What most governments are only waking up to these days, most of these companies have been actively doing for decades. They own and continuously expand gigantic commercial primary datasets, have the ability to snag and retain the brightest talent from each other and from academia, and no investment need is big enough for them – if the prize is right.

Their AI platforms, tool sets and flagship projects (e.g. TensorFlow, M, Siri, Azure ML, Alexa, Watson, Einstein etc.) are already household names that maximize the limelight. They produce significant commercial yields, even if the perceived business use cases are relatively narrow and still just unfolding at this point. B2B sales to laggard big enterprises guarantee sustainability and B2C solutions further enrich their datasets, while their R&D teams compete to conquer the next major AI development milestone. In the US, many leading AI researchers in academia are funded – and for some, at least partially employed – by a major tech company, creating smooth paths to direct commercialization.

The Top 10 tech corporations are dominating the field in both investment and acquisitions when it comes to AI startups, and in most cases ensure through sophisticated attraction schemes that the path to acquisition from day one of a startup's life would be unhindered – even set as a strategic goal, a raison d'être for founders. Aside from efficient scouting and early stage investment or acceleration, a lot of Tech Titan energy is going into open-sourcing their AI frameworks, datasets, coding and learning resources that help bridge the talent gap and attract developers to belong to a single AI ecosystem from the outset.

Million-dollar compensation for ML experts is an offer very few will refuse; this is the daily reality in Silicon Valley that other contestants find hard to beat. That factor alone would be enough to breed the unbridled techno-optimism radiating from the Valley. Also, it may be alluring to some of the less scrupulous talent to have a corporate machinery to clean up after them in case of major mishaps. Fortunately, this same environment was also the breeding ground for the first wave of AI ethicists, standing up in opposition to the mantra we have heard too often: "If it can be built, let us build it anyway: we'll figure out the consequences once we get there."

Regardless of whether you are a fan or a foe of the Silicon Valley version of techno-capitalism, it is useful at this point to objectively define the set of aims and motives Big Tech has in AI development. Fast commercialization will always take precedence over caution due to shareholder pressure and the perceived dictates of a dire, win-all race. Grave mistakes are only weighed against customer/user backlash or punitive regulatory retaliation: the last few years have demonstrated that both these counterbalances are quite temporary and illusory so far. It is especially so on the government regulatory side, due to a combination of technological overreliance on Big Tech for AI government technology, relying on the very focused activities of tech lobbyists, as well as the sore lack of technology acumen within a generation of legislators.

Some corporate pioneers have realized the need for a certain kind of self-regulation, and lately positions have shifted within the majority of AI Industry leaders to saying: "Please regulate us, but do not stifle innovation." To a certain extent, technology industry leaders have recognized the vital business imperative of ensuring the interoperability of their AI platforms. Google, Microsoft, and Amazon have joined forces in an organization called ONNX and there are other emerging similar corporate-driven AI alliances. If nobody else frames the narrative, they will surely try. The business media keeps talking about "the rise of the Chief AI Ethics Officer": based on our on-the-ground experience, titles like this are still very much a nascent phenomenon. We can only hope that the prophecy will be self-fulfilling soon though.

Noting the growing calls for regulation from technology industry leaders proper is a cause for hope, but if the proof of the pudding is in the eating, the preparing of it will still take time. As a first, Microsoft is making its position clear about its conviction that tech companies are responsible even for the unintended/nefarious usage of their AI products (extended responsibility). It is likely that the industry came to realize that growing uncertainty in consumers and the havoc and potential havoc wrought by certain "dark-grey-zone" AI startups – or even some big platforms – in the present Wild West environment is a threat to their business and the industry at large. If a story breaks about the nefarious use of AI by some startup in a risky environment, the public will not point fingers at startups per se, but at AI pack leaders. For a recent apt example, 500 US police stations were accidentally found to be using the services of an under-the-radar facial recognition startup that got away with millions of web-scraped photos of people, very risky data privacy and security practices – all because the field is unregulated, the space is hot, and the need is high. The startup's client list was hacked and made public, and as expected, there were a lot more government agencies and corporate clients on that list.

As hinted at above, gone are the days in Silicon Valley when new startups were bootstrapping themselves and had a higher chance than that of a hobbyhorse at the Epsom derby of becoming a category-defining new giant. Nevertheless, they grow like mushrooms. The **AI startups** now growing up in the shadow of major tech corporations seem to be a mixed lot. What connects them? One major factor is that investors are, for lack of a better term, lemmings when it comes to hype cycles. Let us take a glance at the defining trends.

AI is definitely the hottest investment target out of all frontier technologies. Capital going into AI startups in 2019 set a new record: $18.5 billion in the US only, spread across 1,300 startups. Now, as the arguments of AI transforming lives and business fundamentally are getting more evidence-based, the differences of applied AI business vs. traditional software/SaaS models are becoming more visible.

Leading VC firm Andreesen Horowitz points out that AI businesses are somewhat in-between software and service business models. This is largely due to lower gross margins because of cloud infrastructure bills and manual data handling; scaling is hard because of the unknown time needed to reach accuracy in tricky edge cases; and related defensibility challenges abound on the customer side. These challenges are being figured out on the fly, and the next few years are expected to bring major architectural changes: but the big bets have already been placed.

Consequently, AI investment also brought new startup lifeforms: one phenomenon to address is "instant AI unicorns," a new genre of startups coming to life as necessitated by the very high barriers of entry into the AI game. These are startups reaching the coveted and very media-friendly unicorn status from their birth, traditionally defined as $1 bn valuation.

The Top 100 AI startups show both the targeted Chinese concentration of funding into mega-startups, as well as the historical dominance of US VCs, with 76 of the Top 100 being US-associated. Those on the chart have a total of $12 bn in funding, with about 15% of them being unicorns, and with the previously mentioned Chinese SenseTime and Face++ leading the pack. These two focus on facial recognition tech for governments – not your traditional Silicon Valley vibe, is it? While in the pre-AI era startups tended to do anything for a little publicity, some of the most successful AI startups of the present age consciously choose to operate under the radar.

The other pattern of significance is the sudden proliferation of startups who define themselves as "AI," especially with early-stage ones grossly exaggerating about their technology being AI. This has largely been triggered by the hype affecting all frontier technologies, the sudden popularity of AI investment in abundance, but also from the simplistic mantra of the last few years in the historically Silicon Valley-dominated global startup ecosystem: "Just take X, add AI." Now in all transparency, for many of these startups, AI is a very thin veneer and a buzzword for what they actually do: some industry analysts put the ratio of fake AI startups at 40–60% of the genre. The danger is that it certainly fuels the hype further, increases the chances of a backlash, distorts expectations, and ultimately, may bring about an early AI Winter. We do not intend to underestimate the potential of AI startups though – especially that of those addressing issues of societal significance ("Good AI/AI4Good"), that are largely ignored in the short run by Big Tech, or by VCs seeking the next billion-dollar return.

Having said the above, some early stage "borderline AI" startups already make good business even with smaller-size clients – all they needed was the right high-pain niche. To elucidate the above with a highly anecdotal example, a few years ago the our inboxes were full of spam soliciting outsourced development services in distant lands. Today, we experience the same mindless spamming with “AI startups” offering RPA (robotic process automation) solutions, so that we could “eliminate superfluous employees,” as the marketing tagline goes. RPA is the strongest growing segment of commercialized B2B AI: currently it grows 70% each year. Please juxtapose this contemporary phenomenon with the macro-level prophecies of AI producing a lot more jobs than it takes away and see which sounds more believable in the short run.

As we will elaborate on further for the rest of the chapter, the creeping effects of societal transformation and the dangers associated with it seem to be of a much greater magnitude than your garden-variety commercial disasters, stemming from a too early go-to-market of half-baked AI. With startups, that kind of danger is even more likely than with Big Tech, due to the rules and mantras of the startup religion with buzzwords like “The Hustle,” “Fake it ‘til you make it,” “Move fast and break things,” etc.

The antidote may be what we will call here the ***AI Stewards***, a small (but growing) group of organizations working actively on designing new, consensual frameworks that would enable steering of AI in a beneficial direction, and concentrate on avoiding falling prey to deliberate, or laissez-faire, dystopias. Their resources are meager, their voices are still thin, but their work is crucial, if thankless. Some early adoption signs indicate that it is not hopeless to design ethical frameworks; the challenge is actionability and algorithmic “codability.”

The 2017 Asilomar Principles created by the Future of Life Institute were probably the first of this kind: a declaration setting high-level principles regarding the goal, funding, policy link, culture and cooperation imperative of AI research. The declaration set the first dozen ethical principles around safety, transparency, value alignment, responsibility etc. Most importantly, and almost without precedent, it set its eyes on defining issues preemptively, such as mitigating AI-induced existential risks to civilization, or AI to be directed to serve the common good. These efforts got active participation as well as funding from concerned luminaries like Elon Musk and they have been adopted by some Big Tech companies, as well as the state of California. Safe AI (a subset of these objectives) is the focus of the nonprofit Open AI initiative that also got an initial boost from Musk and has now been adopted by Microsoft with a new focus on commercialization of its research, although with early backers increasingly becoming its critics. IEEE is leading a crowdsourced community effort for setting standards and certifications for the ethical design of autonomous systems. Much of the work in maturing ethical principles is inching ever closer to actual policymaking or is very much state-backed, especially in flourishing Western democracies (e.g. EU, Canada, UK) We think that many more hubs will emerge here with hybrid public-private backing: however, what we need is not more organizations, but more robustness, cooperation, and consolidation.

The problem here seems to be the sheer number of organizations and "scriptures." According to the latest count by the World Economic Forum, there are currently 285 AI ethical principles documents proclaimed by different organizations – some overlapping, some in isolation. Terminologies employed vary, of course, but a high-level consensus is largely visible around AI principles such as beneficence, non-maleficence, autonomy, justice and explicability- with the latter being probably a very novel requisite that AI brought to life.

Critics are already voicing that – similar to the bias found in many ML systems – current declarations are very humanly biased regarding gender, elites vs/ public, global North/South etc. We should be able to use a combination of AI and human research to distill the very essence of these proclamations and make them actionable. That is a necessary prerequisite for any kind of global framework that could put a stop to the current waves of territoriality and fragmentation.

It is by now the focus of many stewards to demonstrate how AI technology can contribute to social development – most prominently, to the UN Sustainable Development Goals (SDGs). McKinsey Global Institute mapped major AI technologies to these sustainable development goals and identified 160 use cases: the strongest ones being in the domains of healthcare, justice/public institutions support, and quality education. More than 75% of the SDGs have been recognized as being augmented with AI technologies, so there is a direct link. As a consequence, global organizations are waking up to the imperative of determining consensual minimums for various fields of life that will be severely impacted by Al (both opportunities and threats). Disturbingly, early efforts seem to be trumped (pardon the pun) by the interests of global powers, their primary constituents and funders. We will see examples of these challenges in later in this book.

What keeps surprising us is the confidence of major AI stakeholders when they frame most AI-driven changes as *inevitable*. As with most lies, this statement – also a logical fallacy – is built on a fraction of the truth. This important grain of truth is that *without major intervention*, a lot of the growing fears of societies worldwide will prove to be justified, if only the logic of the present set of major players prevails. What are these fears and challenges exactly; and most importantly, how can we grow up to the task of reining in humanity's AI future? We will seek answers to that in the remaining parts.

CHAPTER 5

THE PATH OF THORNS: AI ETHICAL CHALLENGES

"Feeling dismay rather than disappointment at this neglect, I immediately sat down and wrote The Scourge of Reason, two volumes, in which I showed that each civilization may choose one of two roads to travel, that is, either fret itself to death, or pet itself to death."

(Stanislaw Lem: Cyberiad)

Triggered by AI, we see this moment in time as an immense opportunity to step back and observe what truly makes us human; what path do we choose to survive and prosper as a global society – and ultimately, as the human race. What is our lowest common ethical denominator? How can ethical AI rules become intrinsic and actionable and feed into regulations? How can we translate these to AI design and programming as directly as possible? What is the optimal ratio between universal norms and cultural differences on that design?

As we have highlighted earlier, at this point the ethical imperative seems to be decoupled from – and overshadowed by – the mass race for AI supremacy. Fortunately, a growing number of experts, policy actors, activists, and informed citizens are asking the hard ethical questions. Recognition of these uncertainties and grave challenges has started permeating the societal subconscious as well, even if slowly. As hinted at before, terminologies vary by organization, but an overarching AI ethics essence is emerging, even if blurred. Here are a few examples to see how terminologies and value focus shift by organization and mission scope.

The EU's AI High-Level Expert Group operates with the principles of human agency and oversight; technical robustness and safety; privacy and data governance; transparency; diversity, non-discrimination and fairness; societal and environmental well-being; and accountability. The OECD's Principles on Artificial Intelligence refer to values-based principles for the responsible stewardship of trustworthy AI:

- AI should benefit people and the planet by driving inclusive growth, sustainable development, and well-being.
- AI systems should be designed in a way that respects the rule of law, human rights, democratic values and diversity, and they should include appropriate safeguards – for example, enabling human intervention where necessary – to ensure a fair and just society.
- There should be transparency and responsible disclosure around AI systems to ensure that people understand AI-based outcomes and can challenge them.
- AI systems must function in a robust, secure, and safe way throughout their life cycles and potential risks should be continually assessed and managed.
- Organizations and individuals developing, deploying, or operating AI systems should be held accountable for their proper functioning in line with the above principles.

For the best counterpart of similar declarations on the other side of the Atlantic, the US Department of Defense (DoD) Joint Artificial Intelligence Center wants defense-focused AI technologies to be responsible, equitable, traceable, reliable, and governable. The DoD also hints at these principles to be evolving and fluid.

There are hundreds more AI ethics declarations. But are expert and elite concerns and expectations shared by citizens? How do high-level principles translate to public opinion, norms and regulation?

In a July 2019 joint study by the World Economic Forum and Ipsos, these public concerns became tangible and "The Concerned" took the lead for the first time in the 27 countries involved. On top of generic AI worries, the majority of the public started demanding restrictions on government use of AI and the respondents expressed an even stronger sentiment for stricter regulation of businesses using AI (close to one-fifth even called for the outright ban of all AI technologies.)

As the public dialogue opens up, the lure is great to say "AI is to blame," but we subconsciously package most of our worries of the digital paradigm under the popular term. That narrative is a surefire way to alienate hardcore AI technologists and researchers. The challenge is instead to delineate and address the worries that are specific to AI and not just technology, "the digital," "the algorithmic," etc. The reason why is the enormous backlog we have reining in previous waves of digital transformation that are necessary building blocks to tackle the complexity of AI. As the expected impact of AI will vary in an asymmetric manner, we also need to be conscious of avoiding sweeping generalizations. For ethics to translate to policy, our scrutiny needs to extend to a defined subtype of AI technology (e.g. decision support systems, or autonomous weapon systems), to AI methodologies (generative adversarial networks or symbolic AI), and to sector-applied AI (e.g. education, recruitment, cybersecurity, or media.)

We now attempt to briefly explore some of the more complex AI ethics areas. Tapping into a variety of leading sources – both representing expert groups and crowdsourced community content – we are trying to operate with emerging terminologies here, and build onto some of the excellent work already in circulation.

Perpetuation of Bias

What are the threats of current AI being "dumb?" One aspect that sheds light on the shortcomings of the present stage of commercialized AI development is the perpetuation and expansion of human bias. Lately, there has been no shortage of headlines denouncing the ingrained frailty of dumb, or "narrow" ML systems.

As described earlier in this book, and as you can also hear through the media almost every day now, the public is being told about racially biased algorithmic denial of loans by banks, or similarly biased and obscure outcomes for "sentencing assistants" in courtrooms. In the world of work, AI software used for human resources screening at top companies was found to be prioritizing male and techie candidates for all kinds of jobs because of the biased initial datasets it has been fed. You have probably also heard about the fledgling AI chatbot that turned fully fledged Nazi in a matter of hours after being launched, due to the composition of its early conversational partners: Twitter trolls having a field day. As already mentioned, there were waves even at higher political circles when AI facial recognition matched US Congress members' headshots against criminal databases, and found a high number of matches – that was also a "color-blind" algorithm's mishap. We have to add a gentle reminder that most of these accidents happened to systems that were already purchased, deployed and live.

It is a separate tragedy that many voters would not see the problem with these outcomes, but we have to point out the specific caveats of this phenomenon. Frankly, Narrow AI can only work with datasets it has been fed, and at its current stage of maturity, this is doubly constrained by both the logic of the technology itself and by the imperfections of the algorithm and data combined.

Without initial ethical design and related screening of prime datasets, coupled with continuous oversight regarding the growing database and outputs, mishaps of a much bigger magnitude are bound to happen. Extrapolating from current trends, the reason why is that AI systems are gradually expected to be given more and more autonomy. One extreme environment is when a decision has to be made fast – faster than the opponent – this is the core question with autonomous weapons systems (AWS). But in environments better known to citizens, reasons can be simpler – and human, all too human. Current applications that are deployed in fast-moving environments are meant to be there to support overworked humans: will an overworked judge overrule an AI decision for someone's parole? Or will we face more defenses like "AI made me do it," across the spectrum of use cases? Ultimately, will machine bias be any better than good old human bias?

Even more importantly, Narrow AI lacks context. Even the most developed conversational AI of our days is mostly an illusion, a mind trick if you like. For an example, a story of what happened in one of the most contested territories of the Middle East: an AI security surveillance system ordered the detainment of an individual who posted a message on social media. The message turned out to be innocent in the end but was interpreted as a terrorist threat by surveillance AI, because of its natural language processing (NLP) and machine translation component not being up to the job. There are many ways this could happen anywhere globally: with an ill-defined threat prediction system, you may soon get a surprise visit from a SWAT team should you play paintball in what is considered a bad neighborhood, and with a friend whom facial recognition has falsely identified as a criminal. Unlike yours, their guns will be very much real.

Explaining the Black Box

Explicability/Explainability is the policy-targeted phrasing of the phenomenon known as the **Black Box,** the aspect that is considered by many as the toughest to crack among the AI ethical challenges. It is a complex, multilayered set of issues. The baseline is this: even if the outcomes of an AI-driven process are intelligible, the steps leading to that decision and output are not. (For a popular culture example, you may also recall 42 being the answer to life, the universe, and everything, without any explanation of the question, but let us leave it aside for now.)

Inscrutability is the most frightening part of many contemporary AI technologies (especially multi-layered neural networks) and this phenomenon is likely to persist in two major forms. On the one hand, a complex AI can only be checked for accuracy etc. by a complex AI. On the other, even AI creators are prey to the Black Box. Yes, this means precisely that it is not only a matter of them being unable to explain an AI's working mechanism, but they themselves have little clue how the AI got to the desired result. This is especially true with deep neural networks and evolutionary algorithms (meaning code written by AI) where the human brain cannot deduce the intermediate steps due to constraints of time and scale. Deep learning networks are made of billons, or trillions of inscrutable connections. It is clear to see that the Black Box is the biggest barrier in front of establishing accountability and an apt regulatory framework when it comes to assuming liability for AI-induced damages.

Explicability is crucial in high-risk, high-stakes systems (e.g. military, clinical, predictive policing) where user trust is a sine qua non. Some AI optimists ask hard questions, though: take clinical applications, for example. If you have cancer and the stakes are high, do you want to be operated on by a human doctor with a 70% success rate or an AI robotic surgeon with 80%? For the foreseeable future, the answer will probably be both: an AI with a human in the loop, or a human aided by AI for certain parts.

While explicability seems to be a cornerstone of most AI ethical principles, the path from principle to policy seems to have inbuilt barriers. Existing attempts at a solution currently seem to have a big gap with the tech forefront of AI systems despite the research efforts going into it (including the one previously mentioned by DARPA). Some of the challenges are "explicability – but to whom?" and how to adjust regulation to address the inherently imperfect nature of ML systems. The explicability aspects that are most relevant to policy are "explain to justify" (establishing liability and responsibility) and "explain to control" (early debugging to prevent large-scale damage). Research and industry are also interested in the aspects of "explain to improve" (streamlined continuous iteration) and "explain to discover" (find hidden, unintended but valuable patterns or expand AI capabilities to less narrow territories).

Cyborgization

It would be hard to overestimate how our behaviors – even our very sense of self – have been utterly transformed by social media or more broadly, the daily ***"onlife"*** in the jungle of cyberspace. When the latter expression was coined by sci-fi authors many decades ago, cyberspace was envisioned as a virtual world you "travel to," with a clear entrance and exit as well as having distinct boundaries separating it from in-real-life (IRL). In our present-day reality, these boundaries have become almost non-existent.

The biggest change of the last decade is that, out of the chaos caused by unrestrained intermingling of our physical and digital selves, the digital assumed control. Its rules, norms, and behavioral patterns started defining the physical as well. As an individual, you may still have very painful ways to opt out of – as well as very complex ways of managing – your digital projection, but it will get harder and harder. Sailing the ocean of cyberspace requires a high level of digital sophistication and constant attention to protective measures: and all that may only provide some semblance of being in control. Regardless of how much of this control or resilience is illusory at the individual level, as a society, we are currently at the mercy of digital storms, mutations, and manipulations over which we have no power at the present stage. On the one hand, the digital ocean is infinite and obeys no one: on the other, key landmarks of our digital world are owned and orchestrated by the very few.

Largely unnoticed, we became cyborgs and our free will is ebbing away – you may have heard this years ago from Musk or Harari. Our arsenal of devices enables us to project ourselves into spaces hitherto unreachable, but we pay the price for that capability, both paying the ferryman with our data, as well as receiving massive quantities of manipulated information that we are unable to process adequately. The digital grid erected around us can be defined as both our circus and our prison. Any effort to change that equation should start with a full understanding of the severity and gravity of this setup.

In our near future, powered by AI, whatever can be hacked, will be hacked. Some of the loudest critics even state that we have already been hacked – and with AI you will be hacked unavoidably, irrevocably, and infinitely. You may already be concerned that your friends and loved ones – as well as growing swaths of society – have been transformed into ideological, political, or commercial automatons: be prepared for this to become exponential, due to the all-pervasiveness of maturing AI technologies.

As brilliant historian turned even more brilliant futurist Harari pointed out, the most dangerous fallacy is – and increasingly will be – the notion that our free will remains intact and we are still masters of our fate, captains of our soul. The kind of mental and emotional resilience to withstand 24/7 onslaught wrought by algorithms that were designed to make you do something, think something, or not do something, only exists in minuscule pockets when it comes to individuals, and is definitely absent at the societal level. Sophisticated tools of manipulation can be built in a matter of months, but it is impossible to orchestrate an "instant cognitive evolution" in humans that counters manipulative forces existing at a mass scale already. (The good news is that, on the other hand, evolution will need hundreds of thousands of years to permanently mold our faces to the duckface expression that became immensely popular on social media in the last few years.)

Barring full digital secession, toolsets enabling resilience and maintaining control are meager and scarce. It is also very difficult to envision these working with a matched sophistication vis-à-vis the pervasive onslaught of AI without the mental capabilities of setting them up, operating them and fine-tuning them. There is no equivalent of antivirus software in the mental realm. The solution is expected by most to emerge in the realm of education, but the pace and scale of change needed to adjust global education systems – that are still very much pre-20th century in spirit, structure, and rigidity – are sorely missing. Furthermore, challenges start at the level of raising awareness, accepting hard truths about ourselves, rewiring behaviors – these are more fundamental than the complexity of an education upgrade. This may well be one of those near-future cases when a major re-humanization effort needs to be defined and driven by an intellectual-social movement, using "humanized AI" for agility, personalization, and scale.

On the current trajectory, we will increasingly rely on the comfort and convenience of AI assistants, guides, coaches, avatars, "daemons": name them as you will. Consequently, we will gradually become prone to living our lives in **autopilot** mode, but with the appearance of our lives becoming smarter and more efficient. The last decade or so has already accumulated enough dystopian signs of how Silicon Valley-originated fads like "lifehacking" ultimately mechanize us while we are chasing higher efficiency and productivity. (Customer service reps talk like chatbots and your friends talk like customer service reps. Sound familiar?) We will become more and more addicted to the comfort of being mechanized, surrounded by choices ever narrowing at the intersection of what data we provide, how AI interprets our needs based on our data input, as well as the interests of AI creators, customers, and advertisers of the broadest possible kind.

Remember the Endless Siege AI Principle from earlier in the book? To give you another example, AI presence does not necessarily mean *more* nudges, but the *most efficient* ones to hit the spot. Instead of today's "dumb" outreach via interstitial ads on YouTube, AI has access to and can optimize suggestions based on hyper-personalized, refined data points fed continuously by yourself, as well as the huge dataset of its user base ("lookalike audiences" etc.). This will all be designed flexibly and for conscious growth in sophistication to fit the objectives of the advertiser.

To be more concrete with what may be considered an illustrative hyperbole: an AI assistant/coach will fuse your calendar, your eating patterns, you mood as it reads it, your health records etc., and will humbly suggest an advertised meal at the exact moment the suggestion is most likely to succeed. It is McMeister's new Fitness Forever Menu, and you may even be thankful for the help. (That until a few years later, when your physician comes to you with eyes cast down.) Once in, suggestions will become regular and more refined, your now daily meal will somehow pop up in your calendar as a recurrence, it will be ordered to be brought to you automatically when you are in a meeting, also you will be reminded how many Meal Points you have gained. Maybe a few meals will be donated to nonprofits every year, just to keep your conscience clear.

Now, picture the above in a cybersecurity context at the macro level. An AI-powered intruder can just pick a target (be it military, government or corporate) infrastructure, order a pizza and wait. It may take some time to find the point of entry, but it will be automated, continuously self-learning and *all it needs is one* vulnerability to exploit. On top of that, the only limit to the number of simultaneous attacks will be the computing infrastructure. This may be considered a highly realistic nightmare scenario of endless war with consequences just as grave – if differently exhibited – as the worst possible outcomes envisioned for thermonuclear war. Ultimately, whatever can be hacked, will be hacked, be it our minds, our infrastructure, our defense systems: "with AI, just set it and forget it."

Ever since the AI scene exploded around 2015, optimistic AI scenarios in various strategy and research papers have emphasized the desirability of "augmented intelligence": AI and humans working in tandem, with systems designed for humans to maintain ultimate control. But how will our changing sense of self at the societal level react to a continuously evolving and expanding AI presence in our lives? Although our psyche gravitates towards incrementalism, this time, based on history, it may also take the extreme shapes of total submission and/or violent backlash due to a sense of inferiority.

Whenever AI ethicists talk about **human agency** and the imperative of supervising AI decisions at critical junctures, one can't help worrying for many reasons. Our top concern is about people who will feel that they are dumber than/inferior to an infallible AI. They may quite happily relinquish their power to intervene, pointing at the AI if questioned and shaking off responsibility. Some may practice overreliance out of laziness, some for lack of bandwidth. Some will even choose to revere or worship AI. At the other extreme, of course, some levels of neoluddism will be inevitable when the out-automated will rise against the machine – as is already in plain view when people try to drive self-driving vehicles off the road in bouts of naked aggression. We believe that we will see both extremes in action soon – the question is whether they become mainstream.

Anthropomorphism

We humans as a race have an unshakeable urge to **anthropomorphize.** If Narrow AI gets to the point of mimicking a limited semblance of empathy, and it has already gained our trust as a virtually omniscient helper, how hard will it be to resist the urge to treat them as emotional companions? Given a certain period of trusted coexistence, how likely will we be to overrule their presumed omniscience and disregard their nudges? How attached will we become to our supercharged Tamagotchis?

Red flags come in various shapes and sizes: mine[6] was kind of trivial and happened just last year. I was driving on my usual route and while I did not need navigation, somehow it stayed on, powered by the most popular AI mapping software. Before I took a very familiar turn, for the first time ever my AI voice assistant said: "Take the second left lane and turn left at (insert fast food chain name here)." I was neither hungry at the time nor have I any appreciation for fast food, so this instruction irritated me to the point of almost tarnishing my otherwise spotless driving safety record. When I regained my composure, I started thinking: what if the next generation of my assistant will have the right data to notice that I must be hungry based on my eating patterns, as well as knowing the type of food I like? Would I appreciate such a sneaky way of hijacking my navigator if the options were personalized or even microtargeted? I found the answer to be a resounding no, but I may well be an outlier. What if I were a senior citizen, living alone and depending on an AI assistant for a constant "human" voice to talk to? What if the assistant added "Yum-yum" to meal nudges in a slightly metallic, but melodious voice?

Anthropomorphism is one of the most dangerous fallacies in AI adoption, but since it is the easiest way to the hearts and wallets of the user, few creators/constructors will disregard it in their development pipeline. Even at the present time, there is a considerable backlash and ethical pressure for AI agents to identify themselves clearly as robots, for example, when making voice calls on behalf of a human user towards another human operator. A few AI creators made pledges to observe these principles. But what if next-generation AI companions will emulate the imago and voice of a favorite fictional character or Disney princess for a child, or that of a deceased spouse for a widower? Who would say no to ordering flowers of remembrance from a local florist advertiser; or the newest magic wand when it comes straight from your secret (imaginary) friend Hermione from Harry Potter?

If we stay truly anthropomorphic, what could mean the end of an era in AI research – and the opening of a new one – is the problem that even the most sophisticated forms of deep neural networks are not yet good enough to play human. They suffer from deep deficiencies that prevent them from replicating some of the most basic functions of the human brain. The answer so far was to build a "bigger brain": but shouldn't it be a *better* brain?

In the depths of AI research labs, the next extended phase will be about the approximation of biological and artificial agents, with human consciousness as the mysterious inspiration – and elusive target. The fringier ones will even attempt mimicking or replicating consciousness. No matter how noble the cause of researchers, for the next few years out in the world where AI is supposed to earn revenue the daily reality for consumers and users may well be significantly more mundane. The best way to imagine anthropomorphized AI agents of the near future is like having an insurance agent acquaintance: give him a few years and in spite of themselves all your friends will have bought a plan, kicking and screaming. AI is capable of doing this 24/7, until they get it done.

Value alignment

It seems that there is a broad consensus that *human values* should be hardcoded as much as is possible into AI development. In moments of chaos, partisanship, and fragmentation, it is only natural to ask: what are these values? What values are innate and what are learned? How much can they be seen as universal? Do we have a moral lowest common denominator that can serve as the backbone of ethical AI design? These are all questions that have been central to moralists, ethicists and philosophers since the dawn of civilization. It is very much welcome that they resurfaced in the 21st century after a long, laissez-faire technocratic lapse that did not treat them as a priority.

A personal confession I[7] have to make at this point is that a decade ago, I was in charge of building a major response to the digital skills shortage gap that was at a high risk of stymieing recovery after the global financial crisis that started in 2007. The desire of the multi-stakeholder coalition and foundation I put together was to create a PPP program that would enable the digital training of millions, imbuing them with the user, business, and coding skills they needed to stay competitive in the Digital Economy. At that time, our vision was bold and much needed, but narrow and flawed: we somehow thought that giving people user skills and access to the internet would automagically create a better society and a better economy. Nobody asked at that time, but if we had been asked, we would probably have answered that technology skills were outside of the moral realm and that people would just bring their own well-established morality into the digital world. So digital literacy never included ethical training; none of us could even fathom the coming backlash of social media becoming a mudwrestling battleground, a repository of obscure, medieval conspiracies, and a hate engine. This was probably my rudest awakening in the years since. I am sure I am not alone in that. The last few years have clearly demonstrated how naive we all were and how the largest missing piece of the puzzle in the 2020s is fusing technology with morality – and enforcing continuous alignment between technological progress and public benefit.

One saying you may have also heard from the music industry was: "It does not matter if a piece of music is good or bad: the only thing that matters is whether people dance to it." That kind of amorality permeates most industries surrounding us today. With

AI designed to make moral decisions, this can't continue this way anymore. But it seems that many technologists still need to be awakened.

In a similar vein to the music industry, for a long time, ethical questions about technology had a single dismissive quip, which says "Technology is neutral." We are sure you have heard it. It wasn't true at that time and is definitely false when it comes to AI. Decision-making, self-learning and self-developing technology cannot be treated as neutral, because:

- AI perpetuates the logic and objectives of the AI creator and its customers;
- AI bases its decisions on data that is almost always flawed in one way or another;
- AI does not get context and morality but still has social and moral impact: the line between amorality and immorality is a very fine one.

To go deeper into value alignment exploration, we are tapping into a useful classification by Swedish professor Virginia Dignum. The three AI ethics dimensions she puts forward to be observed are "Ethics by Design," "Ethics in Design," and "Ethics for Design." Ethics by Design is about imbuing autonomous machines with the capabilities to make moral decisions – so we are building artificial moral agents. Ethics in Design ensures that AI design and development processes take into account ethical and societal implications. Ethics for Design, (that also means Ethics for Designers) is about setting the integrity bar for AI creators together with their accountability mechanisms, like certification and licensing.

Even with the best intentions, AI can't be taught about morality without a thorough understanding of what it actually is, and how gut feelings can be converted into data input. Provided that we get alignment on these values and overcome these coding conundrums: can we hardwire them and create new "digital stone tablets" for machines? Or is it better to be flexible and let machines learn morals by mass interaction with people? The answer is both, but there are drawbacks on both sides.

AI is the first "moral technology" in a sense that it is empowered to make autonomous or semi-autonomous moral decisions, some of them well known, for example, the decision made by an autonomous vehicle whom to choose to hit in case of an unavoidable collision. The elephant in the room is the trouble with high-risk systems: what may be seen as an outlier with self-driving vehicles (albeit a regulatory minimum regarding the assignment of responsibility) is the very core objective of autonomous weapon systems designed to kill. We humans are not flawless at all in this regard; our ability to reconcile our actions with consequences is most certainly imperfect. An imperfect machine's decisions may just become a moral reflection of our imperfect selves. Or will they become new tools and guardians that more accurately describe our moral reality?

For an inspiring example: one of the most interesting new AI development directions, unveiled by Professor Yoshua Bengio of Montreal, constitutes an approach described similarly to the modeling of human thinking by Daniel Kahneman in his book "Thinking Fast and Slow". According to Bengio, "System 1" thinking is like current AI deep learning: optimized for speed, unconscious, non-linguistic/non-contextual, habitual and pattern-driven. His preferred future deep learning would be like Kahneman's "System 2": slow, logical, sequential, conscious, linguistic, optimized for planning, reasoning and adaptation to change. This approach, operating with the primacy of conceptuality, contextuality, and causality currently seem like the most viable to path for both human-level and human-centric AI.

For the public to recognize that we have a choice and for them to support new AI development directions, the first necessary step is to overcome the very comfortable "machine amorality" tenet. If both policymakers and society accept AI's moral implications, then getting a grip on moral machines should become a priority of the highest order. The ethics principles that so many organizations have been laboring on may act as a loose set of "Commandments," a semi-transparent normative framework. It will increase understanding about why AI is different from previous disruptive technology waves and point out the biggest red flags for society. However, realistically, if we stop at this level, ethics principles will only have the power of inciting voluntary adherence to them and *hopefully* may drive the hands of AI creators and policymakers alike. The dangers are too abundant to settle for such an outcome. Through a lot of hard and ambiguous work, the ethics principles should be transformed into a globally harmonized regulatory shell in many parts of the world: we think that we will at last see a "regulatory explosion" in the near future.

Even if nascent policy actions yield some short-term fruit, these efforts are far from enough. "Principles to regulation" may sound difficult but still, is a far easier path than "principles to code." Once this sinks in, we should dedicate a gigantic effort across the domains of research, development, and policy to put human morality as an algorithmic objective front and center for all upcoming AI technologies. This can start (but definitely not be considered to end) with feeding them mass historical data on human moral decisions – and legal/ethical repercussions of them – coupled with real-time interaction and training on human behaviors and choices through unsupervised learning. (It needs to be emphasized that the learning is unsupervised, not the AI's decisions/actions.) This approach may still prove to be a dead end at the current stage of AI technology, but no need to delay: it will become fruitful at the next AI milestone. Regardless of the present viability of yielding an ultimate solution, an ever-widening, controlled mass interaction between a large group of humans and AI agents, focused on a sophisticated understanding of morality for AI decision-making is a must. From the aspect of citizen well-being, this public-interest effort deserves a similar level of attention and resources to some of the military or commercially focused AI efforts that have been going on for decades.

Parallel to the complex concept of "Moral AI", an emerging field is "Compassionate AI". This should be mentioned as a stepping-stone to the popularization of what objectives AI should serve, what is or should be its raison d'être. Compassionate AI is broadly defined as designing AI with a focus on alleviating human suffering and focusing on social impact at both the individual and the societal level. As a long-term objective, hardwiring compassion seems even harder, though, than instilling machines with basic morality, as it expects machines to understand, empathize, and emote genuine human love, kindness, and concern. But we can start small and redirect the nascent field of emotive AI (the subdomain specializing in AI interpreting human emotions) into compassionate AI as the teleological direction. To oversimplify it: we can choose whether we accept/prioritize emotive AI that scans you in a grocery store and determines pricing based on your emotional state, or if we should focus our resources on the vast potential of compassionate AI agents?

In Japan, an aging society that is resistant to the idea of mass immigration to tackle the eldercare needs of the country, Compassionate AI, especially home care robots, is gaining ground. Efforts like the Japanese ones are the conceptual antithesis to the Skynet/ Terminator vision implanted in our brain by Hollywood. Compassionate AI advocates see evolving AI agents more like a "companion race" to humanity, be they nurses, mates, chaperones etc. Can this go wrong? Of course, in so many ways. But what if regulators worldwide could go bold, and prescribe and incentivize compassionate AI as the preferred direction for society and as the objective for AI creators to follow? Wouldn't we be much better off already if – instead of lagging behind industrial interests in a laissez-faire way – the role of AI were defined in the societal subconscious as the one closest to the most fundamental needs of people?

Data Ethics

Facebook has 52,000 consumer ad subcategory labels to serve advertisers and 190 personal major data points on you as a user (you can actually check the latter for your own data profile). How are they being collected and how accurate are they? If one looks up how one's political views were labeled by the No. 1 social media platform, many readers will certainly disagree with their classification. Who is infallible in that regard? All content served up to you for agreement – or, the even more lucrative fervent disagreement – is based on that one final piece of AI-based judgment that put you in a specific basket at a certain point. Can we write to customer service asking to be reclassified or do we have to start liking pages with opposing political views out of spite, to nudge the AI to revisit our case?

Actually, do major Tech companies listen in on your conversations? Anecdotally, many Facebook and Amazon users we know have a story or two to tell about getting product ads about things they have only raised in a live face-to-face conversation with another person. In 2019, Consumer Reports found that 43 percent of Americans who own a smartphone believed that their devices were listening to them. Eavesdropping has not yet been officially proven, and companies keep denying such snooping, especially in conversations with the regulator. The emerging consensus among corporate advertising data wonks is that eavesdropping continuously on audio conversations of users would require a data storage capacity one magnitude higher than is currently available for Big Tech. (Does that then imply that it is only a matter of time?)

The other rebuke (that you may like even less) is that they do not ever need to listen to your conversations to find out how to target you. By means of extensive partnerships, Big Tech has such depth, quantity, and variety of demographic and psychographic data that the ad choice you see may only be based on the assumption of your consumer habits being a mirror of a large number of your "digital consumer relatives," i.e. people with significant behavioral overlaps with yours. It is probably a notch less creepy than constant surveillance but absolutely no reason not to feel alarmed. Individual choices become more illusory as AI will match us up with our data profile and not the other way around. In the political realm, it is even more obvious that the tribe you will be constantly nudged to join according to the dictates of the political advertiser's interests, will force you to adopt all the paraphernalia associated with your tribe, be they ideological products, or products forced on you by sophisticated, AI-driven peer pressure and deliberate tribalization.

In today's equation, individual data privacy, consent for usage and ultimately, data ownership, are all lagging behind our daily realities. Remedying that unsustainable setup is an imperative both at the regulatory level, and at the individual level. Just as we needed people with driver's licenses and lots of new car mechanics at the dawn of the combustion engine age, we need morally conscious AI developers as well as data-informed and data-skilled citizens on a mass scale in the AI Age.

CHAPTER 6

AI & DEMOCRACY: IN MACHINES WE TRUST?[8]

What will the impact of AI disruption be on democracies? Elon University and Pew Research Center's 2020 study asked experts: about half predicted that humans will use technology to weaken democracy over the next decade, with their concerns grounded in the trends of our present. The key themes are that AI will further increase power inequalities, with systems exponentially exploiting the data-illiteracy and AI-illiteracy of the public. As a similar level threat, public trust in institutions, and in each other, is expected to ebb because of the combined effects of disinformation and the confusion it creates, the decline of fact-based, objective journalism, and society's inability to keep up with the speed, scope, and impact of manipulation technologies.

On a more hopeful note, about a third of the experts interviewed believe that we will use AI to strengthen democracy. AI could be an excellent tool for raising awareness, helping human systems evolve and adapt, empowering responsible governments and social movements, and assisting in upgrading democratic frameworks while enshrining values. We think that both trends and trajectories are likely to be strengthened, but short-term trends are primarily benefitting dubious actors. Why?

Digital technologies have opened new channels of communication and coalition-building that allow for direct interaction between political leaders and citizens. They have also created a space for the expression of political ideas that might otherwise not find their way to the political debate. Digitalization has lowered the barriers for citizens to engage in nationwide and global political conversations, and amplified these conversations. The digital brought to life a new world with flatter, horizontal structures whereas historically representative democracy has functioned as a hierarchical vertical process (and is therefore deeply challenged as of now). This transformation has facilitated access to the political market for new actors, sometimes with a strong and often unscrupulous communications and digital strategy.

While serving as an alternative outlet for free speech in countries where this right is not a given, digital technologies also bring about new information challenges, namely the skyrocketing scale and speed of development of a persuasion/manipulation architecture that is derived from the current engagement model of social media, fueling the "attention economy." Unwanted side effects of online anonymity, zero-cost publishing, and content retransmission are that they favor the propagation of political messages, hate speech, and extremist and polarizing ideas. **AI Algorithmic content curation**, whereby AI recommendation engines, search engines and newsfeeds prioritize content based on each user's profile, have created echo chambers that push online citizens further into their ideological bubbles and away from multi-faceted analysis. AI can block the construction of public debate. AI-driven technologies are designed to catch and capture our attention to make advertising space valuable - to maximize the time spend, the virality and the attention the ads get.

AI-influenced public dialogue is no longer science fiction: it is now. While narrow in its focus at this point, it is already omnipresent, especially when it comes to rewiring our minds and the remnants of our Digital Commons. As it powers all the major digital social platforms that we are using, AI is now heavily influencing our information, conversations, choices as citizens and consumers, our relationships – our very lives. Therefore, we need to understand and analyze how AI technologies are being designed to exploit our human vulnerabilities as individuals and societies for economic and political gain – or pure disruption. The damage is already done and transformation brought about by AI putting manipulation on steroids has fundamentally shaken the public dialogue and democratic prerequisites. The time to act is now.

AI is a double-edged sword, with both threats and benefits to democracies worldwide. How can democratic societies harness this technological force of unprecedented power for good and use it to the public advantage? The imperative is not just fast adaptation of agile policymaking but also setting a shared vision of clear choices of what kind of AI is important to fuel the public dialogue and support democratic institutions. Policy, industry, and citizens should work to fulfill "Good AI's" promise for inclusive societies and democratic guardians: that could potentially yield an unforeseen level of public engagement. Good AI is designed and implemented based on the public's values, as articulated through a deliberative and inclusive dialogue between experts and citizens –AI is ideal to power such dialogue and collective intelligence at the same time. This is what should drive policy's and society's expectations of new AI tech towards AI constructors. The AI solutions we choose should be less "Artificial", but bring "Augmented Intelligence": a sustainable, humanistic vision that empowers both digital citizenship as well as radical speed and efficiency boosts for democratic governments to craft connected policy.

Trust Dismantled

Trust has been eroding globally for a long time, but the decline has accelerated and been further disrupted by digital waves of increasing intensity. We are at a precarious point in history when our fundamental institutions – government, public institutions, the media, corporations – are not seen overall as trustworthy by a majority around the world.[9] The public perceives the most important drivers of trustworthiness to be reliability, transparency, and responsible behavior: there is clearly a gap as of now.

Digital communications and social media behavioral phenomena have aggravated the "perils of perception": perceptions of trust are often out of line with reality. With Millennials slowly taking charge, societal trust is significantly lower than with preceding generations:[10] low-trust environments are fertile ground for disinformation as the public loses confidence in impartial arbiters of a common set of truths.

Journalists are now one of the Top 3 least trusted professions.[11] The media made several mistakes fighting economic decline and the decimation of the journalist workforce including: the pressure of optimizing content for social media, failing ad-based business models, 24-hour "Breaking News" attention desperation, ethical decline, and in some cases unbound partisanship. One-third of the public trusts media less than they did five years ago; more than 6 in 10 think that online news sources contain a "great deal" or "fair amount" of disinformation.[12]

These are complex times indeed. Big social media platforms are the primary gatekeepers for news outlets: even emerging media business models cause media to possibly increase the bottom line but ultimately to forfeit their destiny to the social media aggregator. Platforms accidentally gained an overwhelming amount of power and discretion to decide what content reaches us, and what counts as harmful: their business interests are not easily bridgeable with the public interest of constructive dialogue.

Digitalization has led to the atrophying of online public dialogue and to the unexpected realization that "connectedness disconnects." In the social media world, a concentrated, loud minority opinion of a few percent can create the illusion of being the aggressive majority, especially if those who want that opinion spread magnify it with paid ads and posts from fake agents, both humans and bots. Both the disappearing middle and waning empathy are detrimental to democracy: we are being pushed into corners/camps owing to a combination of exploitative tech (e.g. subpar AI labeling of our views, recommendation engines keeping us in our echo chambers) and psychological weaknesses such as vilification of dissent, or groupthink. Understanding and engaging the silent, confused, fearful "bystander" majority, who accidentally handed the town square to extremists, is key.

AI-driven disinformation

The last few years have brought the overuse of the term "fake news" to the point where it is devoid of any meaning, so we will use **disinformation** consistently. The public's awareness and frustration with online manipulation is clearly demonstrable. Four in five global citizens admit to having been exposed to disinformation, and four in ten (44%) admit to being duped by it. In the US alone, ramping up to the 2020 elections, politically relevant disinformation was found to have reached more than 158 million estimated views, enough to reach every reported registered voter at least once. Some 86 million views of disinformation in fall 2019 shows that the amount tripled vs. the preceding three months.[13] However, few can agree about who should police the digital commons and determine what is fake.[14] The extreme challenge here is that disinformation often involves legal but manipulated content. Unlike terrorism, violent extremism, or child pornography – universally considered unprotected speech – platform action on disinformation is not as clear cut: understanding the distinction between legal and illegal speech is important.

Strong majorities of people support all forms of action to resist online disinformation, save for extreme government censorship. Some 75% say that social media has too much power and platforms are responsible for distrust. Public opinion is inconsistent, however, on how to deal with this matter: six in ten global citizens say that it is acceptable to temporarily cut off access to social media platforms during times of crisis to prevent the spread of misinformation. However, if the question is framed differently, six out of ten also agree that temporarily cutting off social media platforms at times of crisis is unacceptable because social media is the primary source of news and information for many people.[15] Probably this is the most illustrative example of why society is in a state of mental paralysis when it comes to concerted action aimed at risk mitigation.

Propaganda and mass manipulation are far from new, but boosted with AI, they can be even more detrimental, to the point where they can threaten democratic collapse. (Currently, 70 states are engaged in AI-powered disinformation campaigns; also, evidence exists of 26 authoritarian states harnessing social channels to suppress freedom of expression, discredit critics, and drown out political dissent.)[16] For bad actors trying to flood our minds with propaganda, AI prowess projects an image of competence, while democracies meanwhile struggle to come up with the right answers that combine a humanistic vision, citizen protections, just-in-time agility, international frameworks, and the necessary concentration of resources to be up to the challenge.

In the context of democratic threats, AI algorithms provide unparalleled capabilities in *sourcing/scraping, processing and analyzing* mass data from social media. AI's ability to *hyper-target and deliver* misinformation based on data we share unknowingly, is already being exploited extensively. AI tech is used to *impersonate* real people. It can create mass fake profiles, show fake power and influence, and sway public opinion. The impact is widespread and access is easy: an officially retired powerful fake commenter AI engine (originally by OpenAI) that can imitate the style of any person, reappeared in the public domain, remade by two developers with an investment of a mere $5,000.

The disinformation arsenal is increasing to higher levels of sophistication every minute, every day. Facebook recently announced that they are deleting 1,000,000 newly created fake accounts *every single day.* With AI becoming core to digital manipulation, one emblematic weapon of disinformation is deepfake technology – the most dangerous toolset of the present. Deepfakes are AI-manipulated audiovisual pieces created to look legitimate and can be AI-produced in myriad different versions, hyper-targeted at the individual: even current AI technology makes the barriers to deepfake very low. Fighting them will be a very resource-intensive race. Catching a deepfake only allows the AI network to produce a new one with better quality in a matter of seconds. The prevalence of deepfakes also enables the "liar's dividend," where a politician can claim that a damaging real video is a "deepfake," further obliterating the meaning of evidence and factuality.

For now, most deepfakes take the shape of AI-powered video of somebody saying (and soon, doing) practically anything the deepfaker wants them to say. Current versions may seem oddly artificial to the untrained eye but they are improving at lightning speed. Even if it's an obvious fake, due to tribal instincts and the sensationalism that prevails over common sense, they are shared ad nauseam over social media. The first amateurish deepfakes were pornographic or intended as humorous: but what's next?

Imagine a deepfake of your country leader announcing a state of war and imminent invasion; or a presidential candidate admitting on video to operating a pedophile ring. What damage can these wreak before order is reinstated and the video is purged? Now also imagine that a dedicated AI can produce, tweak and hypertarget a limitless number of deepfakes, only constrained by processing resources. Hypertargeting in this context means that the message you get from the deepfake video is *individually* tailored to the data available of your likes, dislikes, sensitivities etc. This is the next evolutionary step after the tactics successfully used among others by the 2016 Trump campaign. Its social media arm operated with a wide variety of advertisement versions precisely targeted at algorithmically defined and identified niche groups of Facebook users based on hundreds or thousands of data points per individual.

There are promising, albeit temporary, countermeasures being developed by national security agencies and social media giants, but even the most optimistic outlooks point at a strategic arena of an AI arms race unfolding regarding deepfakes. First, there will always be unmonitored channels where deepfakes can further radicalize extremists, even if wider spread is being contained. Second, lacking appropriate technology breakthroughs, chances are that countermeasures will follow the pattern of breach, containment, patching and upgrading the defense engine. But with a robust AI disinformation machinery backed by a foreign country, a single breach can cause detrimental impact in minutes or hours as it spreads exponentially.

A minuscule slice of this broader **"Impersonation AI"** phenomenon is the recent discovery for the public that GANs (Generative Adversarial Networks, one of the current edges in Machine Learning) can produce limitless new digital headshots based on a database of actual human faces (perhaps scraped from Facebook or the free Google or Amazon Photos databases). These artificially generated faces – belonging to non-existent people – are, by and large, very convincing to the naked eye, and definitely so when used as low-res social media profile pics for fake profiles. This new capability is a bane for stock photo companies, and a boon for those who found their fees excessive. But the dangers here are tremendous. Before this point, fake profiles and disinformation bots have been relatively easy to discover by a simple but manual Google image search by the concerned user, and through applying similar AI-supported fake identification tactics at the macro level. With a few clicks, you were able to figure out that the profile picture of the Kansas senior citizen bashing your political views on Facebook actually belongs to a retired music teacher in Poland, whose headshot was featured in local Polish news for her ardent work with talented pupils.

Now with AI CGI profile pics, AI can generate and populate an unlimited number of new, believable robo-profiles that act in a coordinated way. Imagine this happening at a time of ruthless political campaigns when support of a much greater magnitude than the reality needs to be demonstrated; or with major petitions signed on social media to sway political decision-making. With some major societal debates like climate change, major social media companies have already demonstrated that at least 25% of posts and comments are bot-generated. Also in a more financially driven application, the commercial value can be priceless at the launch of a new product that needs a million (fake) reviews to break through.

For the past few years, the political elite and intelligentsia perceived growing tribalism as the biggest threat to the trust fabric of society, but we are convinced that the next frontier will be **digital solipsism**. In this state, AI-powered hypertargeting will completely disassociate fact from conviction, rupturing the very fabric of our individually and societally perceived realities: shedding our skins as social animals, we may well become islands of one.

Digital communications algorithms were designed to optimize for growth and therefore they favor interaction types that oversimplify, taunt, divide, magnify and conquer. That coupled with our personal data map getting more nuanced and sophisticated yields a social phenomenon that is increasingly threatening. If you and your best friend on social media differ only by 10% in your major opinions, registered as such in your data points, in the current algorithmic and behavioral setup you stand a high chance of being focused on what divides you. Consequently, it is easy to slide into a spiral of disagreements that can easily lead to the end of that friendship. On the other hand, with your information sources hyper-personalized, you may have a significant gap even with the fact base of your 10% different friend; will you even be able to share mental models, values and facts with a person whose opinion data map is only 30% different? To our knowledge, with the current trajectory unchanged, this digital media solipsism phenomenon is the most dangerous short-term entropic threat not just to democracy, but to civilization.

Algorithms for the Public Interest?

There is a growing consensus regarding the identification of a wide variety of **challenges and threats to public dialogue** and social cohesion by AI-powered technologies. Detection of AI-based hate speech on social media is reported to be racially biased. [17] Recommendation engines and videos on autoplay are claimed to take citizens down a rabbit hole of radicalization.[18] Data ignorance increases people's vulnerability to AI-powered exploitation, and mass manipulation and adds to their sense of powerlessness. Armies of underpaid human moderators are working under contract to label data and to help bridge the gap of AI's weaknesses in detecting and deleting disinformation and hate speech. Given the meager incentives, will they choose to override a machine's decision? Basic legal frameworks are lacking when it comes to establishing distributed legal responsibility for disinformation campaigns; there are many stakeholders in the digital value chain – including tech providers, developers, advertisers, the platform, the medium, moderators, and the user: it is unclear who is liable if something goes wrong.

Citizens worldwide react to these perceived threats with a mixture of confusion and concern: 40% globally are worried about AI use, with the concerned taking the lead: one in five even wants to ban AI outright. More respondents agree than disagree that governments' and companies' use of AI should be more strictly controlled.[19] Majority populations globally, especially in the developed world, perceive AI algorithms we use every day to be biased. Due to a lack of transparency, naysayers perceive algorithms to be exploitative by design and they believe there is no human element in decision-making.[20]

While they are fulfilling a semblance of a "global town square" function, many people feel that AI-powered social media algorithms are currently not in line with the public interest. Policy's struggles with understanding and regulating social media are far from being resolved. With legislature in limbo, AI pack leader companies stick to their own Terms of Use, lacking adequate regulatory frameworks: some of them say they self-regulate, some call for regulation from others. Current challenges range through policing hate speech (e.g. Facebook's alleged role in Myanmar genocide)[21]; data abuse combined with hyper-targeted (political) advertising (e.g. Cambridge Analytica case and many more); the threat of surveillance capitalism and obscure data monetization; universality of values vs. different cultural norms; establishing accountability in misinformation campaigns; and general transparency and explicability of algorithmic decisions.

The test of our time, largely driven by the rules of the Algorithmic Economy, is how to increase understanding of AI, connectedness and consent of policymakers, experts and researchers, tech business, and the public in concert. As a first step, the last few years have been marked by simultaneous efforts to create consensual **AI ethics frameworks**, from both the top down and the bottom up. As referenced earlier, hundreds of AI ethics code proclamations co-exist or compete in the world. How can we integrate, simplify, and make them policy-ready, especially the ones that affect public dialogue and society, given that the general public perceives this domain as the most tangibly dangerous?

Ethical guidelines play an important role in increasing the understanding of policymakers and citizens alike regarding AI's unique characteristics: both its opportunities and threats. However, it seems like Policy 1.0 is in endless catch-up mode with AI-powered Tech 5.0, and the agile and fluid policymaking the latter demands. AI requires an added **anticipatory** lens: new kinds of regulatory fast tracks and fast improvement cycles adopted from the Tech world – all that while continuously ensuring citizen alignment.

New frontiers: Upgrading democracies for the AI Age

In the past, humanity has successfully come together to co-create laws of our oceans, war and even of space. As of now, we haven't done that for the internet, global public dialogue or AI. Article 19 of the Universal Declaration of Human Rights was not prepared for the age of disinformation: consequently, we struggle with both the philosophical and legal frameworks to address the radically changed disinformation landscape that became supercharged by AI technologies. We also need to co-design new checks and balances that reflect this new digital-first reality.

The dawn of the AI Age should be a stepping-stone to deep reflection for policymakers on the impact of a drastically changed environment on political philosophy and on the necessity of remaking democracy. Anti-establishment sentiment and anti-political class exhaustion vs. perceived AI efficiency reached such levels that an EU study found that on average, 33% of European citizens would allow AI to make important decisions about running their country.[22] The public may be ill-informed about AI but their concerns need to be addressed properly.

The very essence of democracy is in turmoil: falling trust, a growing empathy gap, tribalization, purposelessness, diluted responsibility, emotional reactions to complexity and abundant cognitive biases are all significant dangers. If adaptation is unguided and unmediated, the chances are that our future democracies, while becoming more and more "direct," will very much resemble the increasingly disturbing face of social media. If fully utilized strategically, **AI's promise** for democratic societies is **drastic augmentation in public-interest efficiency, engagement, and insight.** However, for responsible citizens and policy actors worldwide, conscious or laissez-faire disruption can only be countered with deep understanding, modernized deliberation, connected policy and concerted action to **rebuild the global town square for the 21st century.**

Forging the path ahead

The last few years of budding activities by several democratic communities on laying down AI ethical frameworks have been spent with establishing a strong link between AI and **international human rights law.** The next wave should be about moving beyond that in vision, speed, focus, pragmatism, and actionability, but using that framework as an established fundamental and point of reference seems fit.

Most of the impact of AI misuse on public dialogue does not stop at borders: and in many cases, it is designed to primarily impact other countries in a harmful way. Therefore, all interventions – especially ones that require rethinking our models – must prioritize creating and upgrading **international and global frameworks**, the fundamentals of which can be extensively upgraded versions of existing structures protecting human rights. The UN (e.g. UN Digital Cooperation Group) and other multilateral organizations should be spinning off new, fast-moving, cooperative, and actionable structures that ensure the creation of such frameworks.

Policy answers also need to be determined, deliberative and nuanced: there are lots of pent-up emotions that can only be overcome if the new structures **combine public-private models with constant, upgraded citizen dialogue.** If we do not come together in new frameworks that are built on trust, mutual recognition of public good, and a joint harmonization of diverging interests, we will end up in an endless game of "catch me if you can" leading to a pervasive lack of accountability and sustainability in the long run. We made a conscious effort to think much broader than narrowly defined regulation and to recommend framework structures that enable the above objectives. We believe that a long-term solution must encompass a novel **coordinated omni-stakeholder frame-**

work. It should bring together citizen movements, public interest and public accountability that are recognized by the tech industry, and tech business interests facilitated by a policy domain becoming much more agile, inclusive and pre-emptive.

Most experts agree that the current relationship between policy and Tech is ever more strained and hampered by mutual distrust. The **speed of AI-induced changes** may often trigger primarily punitive policy responses in a domain where the current regulatory toolset only really allows you to regulate situations that were valid years ago. The world of policy must focus on consciously adopting as much as possible from the modus operandi of how technology is being developed, updated and deployed.

Domain experts cite three potential targets of countering/regulating disinformation: manipulative **a**ctors, deceptive **b**ehavior, hateful **c**ontent **(ABC model).** Free speech advocates (more and more in line with social media corporate lobbyists, we need to add) say that C should be left completely untouched: it is certainly a sensitive point as a cornerstone of democratic models. Therefore, they say, policy interventions should only focus on removing illegal content, and should apply different principles to harmful but legal content. AI could be a perfect tool in identifying behavioral patterns of ***amplification,*** when a piece of harmful content is made viral, orchestrated by bots, ads, agents and by exploiting the misled: this is where deletion/retaliation should occur. A recent proposal by the French government is a good example of not regulating content, but rather establishing a dedicated regulatory authority on digital communications platforms that will oversee the continuous observation of community standards on clarity, consistency, and fairness.

In this line, special attention should be paid to the role that social media has on increasing disinformation levels. Some initiatives to counter this problem have already been put into place. At national and international scales we are bearing witness to renewed efforts to rethink and revisit the current rules and legislation applied to AI within the context of social media governance. An inspiring example that was born out of the horror and momentum felt after the 2019 Christchurch terrorist attack, the Helen Clark Foundation launched a new project named The Christchurch Principles. This set of 10 principles, inspired by human rights and democratic theory, aimed to reach a common understanding and encourage social media companies, civil society, and government to take political action and make decisions on what social media governance should look like. An initiative like this serves as a reminder that in order to see a positive change in the current AI ethical frameworks, national governments will need to consult a wide range of actors. This could ensure that our new AI parameters work for us, as a public and as individuals, and return us to the promise of the democratic space that the internet originally offered.

Experts and the public are agreed that there is a dire and urgent need for AI policy intervention in the realm of digital public dialogue, especially concerning AI-powered disinformation. We have been consciously trying to set up a rough framework to help policymakers distinguish between three categories where we think different approaches are needed to tackle AI challenges, as below:

- **Category I: Dousing Fires**

In many ways, we have already been hacked successfully. Some destructive AI impact is already out there and needs urgent and drastic intervention. Deepfakes, impersonation AI tech, and state-backed disinformation campaigns are prime examples that call for fast action, involving a combination of technological and awareness-raising actions. In this category, optimally our focus should be on types of interventions that will take less than a year to show results in the list of thematic recommendations (described further below).

- **Category II: Managing AI transformation**

Some AI public dialogue impact (media paradigm shift, facial recognition, recommendation systems, online behavioral norms)) is creeping but deeply transformational already. They could become irreversible in the mid-run without adequate policy answers in place (or in the pipeline as of now); conscious experimentation and expanding policy pilots must be the answer to that. We must infuse this effort with normative approaches: a conscious understanding of digital-induced behavioral change in democratic societies and a trust-building objective. This effort should have a twin social and technological focus and encompass both trustworthy AI, as well as the trustworthiness of the broadest set of human stakeholders behind AI. Ethical frameworks should be designed to translate smoothly into principled policy action. As stepping stones, the Tech policy toolset needs to prioritize **codes of conduct, standardization and certification approaches** that build the foundation for trust and omni-stakeholder buy-in. As an example, in 2018, the European Commission persuaded four online platforms and a few advertising industry trade associations to agree on a self-regulatory Code of Practice to address online disinformation. From the companies' June 2019 reports, the EC found that progress was made on transparency of political advertising,

> "that actions taken against abusive use of bots and fake accounts have helped to detect, debunk and close down manipulation activities targeting the elections Google, Facebook and Twitter improved the scrutiny of ad placements to limit malicious click-baiting practices and reduce advertising revenues for purveyors of disinformation, for instance by taking down ads and closing ad accounts due to deceptive or inauthentic behavior."[23]

- **Category III: Changing the Paradigm**

For some of the impact caused by AI technologies, we need to prepare **pre-emptively** with policy answers that are more **normative** in nature, and this will often involve creating **new institutions and setting a new vision of digital democracy**. Some of our democratic architecture needs fundamental rewiring when it comes to existing models and this redesign process needs to be omni-stakeholder and diversified. This effort needs to be strongly tied to the technological development milestones foreseen, e.g. current structures can't keep up even with Narrow AI that measures development in hours and days: preparing for the coming of AGI (Artificial General Intelligence) with a pre-emptive policy framework is essential.

While the work seems enormous and sometimes isolated, there are many efforts underway – government initiatives, bills proposed, and multi-stakeholder or corporate initiatives to address artificial intelligence in the context of disinformation.

Thematic Policy Recommendations

Empowering Digital Citizens

Education and awareness play a key role in tackling the challenges of empowering citizens for the digital world. Existing vast research on cognitive biases – especially tech-induced ones – should be guiding the hands of policymakers. In the AI Age, empowering independent thinkers as conscious citizens is by far the most important objective of democratic education. In this battle, AI-supported disinformation needs to be fundamentally understood, and rejected, by each citizen. While policy can set the rules, the final frontier is the human mind, so individuals needs to build up a double resilience that pays equal attention to providing protection from digital manipulation and to building self-awareness to counter cognitive threats. To ensure a free, publicly vocal and informed citizenry, this should take shape as a **"Live Digital Citizen Curriculum"** with topics including empathy, critical thinking, media literacy, data literacy, AI literacy, and security literacy. This should be built into public education systems as well as being pushed into adult education and mass education/awareness/communication programs. To ensure the importance of this effort is recognized, global education measurement systems should also build it in (e.g. PISA). Several nonprofits and universities have developed AI citizen education curricula lately that could serve as a great starting point.

Rewiring Media

Quality media is a cornerstone of democratic interest and needs to be supported as such – but support does not need to be unconditional. To earn back trust, the bar needs to be set higher. Both traditional and social media need to be in line with the public's expectations for trustworthiness: we need new **truth-seeking institutions**, especially when it comes to redesigning media. New exemplary online behaviors should originate from unbiased individual experts/organizations that hold a majority of society's trust and will set new norms of responsible behavior in the global digital town square. Combined efforts are needed for government regulation and a global ethics consensus, and for a fundamental rethinking of the technological and economic incentives that led to some media outlets breaking ethics rules.

Fact-checking Boost

Some disinformation experts state that "fact-checking is the new journalism." Live, AI-powered fact-checking should be a global multi-stakeholder effort with robust public/private investment and with the aim of integrating and elevating a multitude of isolated fact-checking efforts worldwide. However, despite public investment, direct government control should be minimized. Public interest AI development should prioritize finding a way to fact-check fast-breaking stories where rumors are rampant and substantiation thin, as this is essential in crisis situations. The signatories to the recent UNGA Christchurch Call have made the latter a priority, and there is also some dynamism in the EU regarding a sustainable European fact-checking network.

Rebuilding the Democratic Core

While a lot of research has been going into dissecting populism and extremism lately, the objective should very much refocus on understanding the disillusioned majority and actively rebuilding the democratic core. For AI to serve the public dialogue we must rethink the rules (and regulatory requirements) to algorithmically reward trust and constructiveness and to emphasize the spectrum of opinions instead of the extremes. With the speed of change in public opinion, this can only be done efficiently with **AI-powered citizen intelligence:** real-time, multi-data source, public-facing citizen intelligence platforms that analyze and track the whole spectrum of public dialogue, with the aim of finding common denominators for policy formulation. AI can also be used to bridge quantitative reach and qualitative depth for mass citizen digital feedback and conversations. Boosting nascent, currently underfunded civic tech efforts towards fostering public deliberation online via institutionalized processes is therefore key. There are convincing emerging attempts within some governments to hold national discussions on AI that will need AI to be used as a tool for good. These should be paired with a **public interest fact base:** shared, accessible, independent, and reframed as a public service. Some core elements of how Wikipedia has been built and some of its functions may be considered as building blocks for the solution, but in a more dynamic and agile manner.

Public Interest Technology Stack

We should not shy away from fundamentally rethinking the digital town square. Multi-stakeholder efforts should establish public interest tech labs that produce public dialogue-enhancing solutions (platforms, apps, plugins etc.) with the speed of startups, that have access to funding and create interoperable tech. Societies need to envision and experiment with a new "Public Interest Tech" category at that the international level that could take shape as, for example:

- Publicly funded, publicly accountable, nonprofit public dialogue platforms with rewired incentive and reward structures (e.g. similar to the privacy-first Signal model in the messaging world – funded by a nonprofit foundation). In a promising move, Twitter recently announced the intention to actively support the creation/elevation of an **open, decentralized social network standard**.

- **Joint venture** between responsible Tech platforms and governments to co-create a new breed of public benefit forum that is not ad-based, has vetted information, where citizens own their data and algorithms are wired to foster consensus, civility and harmony. It is possible under the umbrella of the UN or another global organization: it should have public-private funding but be shielded from direct government influence (forebearers include the BBC [funded by license fee in the UK, but with a commercial wing for syndication], C-SPAN [entirely funded by the cable TV industry as a public service], or PBS).
- A regulatory requirement for Tech platforms to create **"public benefit sandboxes"** as part of their platforms for open Public Dialogue Tech experimentation.

Good AI Policy for Public Dialogue

To get practical on AI ethics:

- Aggregate AI ethics codes: Using both AI and human curation methods, the next step should be to build a single overarching **global framework that hosts an actionable, policy-ready AI ethics code** that sets a clear and actionable vision to be adopted by global institutions.
- Simultaneously, ethical accountability must be addressed not just at the macro level – which will take significant time – but can be fast-tracked at the individual/professional level. Since many professions have existing, strict **ethical codes**

of conduct with elaborate sanctioning systems enforced by independent professional bodies (e.g. doctors or attorneys in the US), these can be remodeled with relative ease and speed to cover the whole value chain of AI development and operation professions (consciously going beyond tech only), also extending to media.

Prioritize AI Development Directions

Policy actors should feel emboldened to pick, steer, and mandate certain beneficial AI development directions, as well as to monitor and scrutinize high-risk ones. There are nascent technologies and AI research directions that prioritize a human-centric direction and aim at cracking the consciousness conundrum as well as prioritizing symbiotic, but closely guarded machine-human coexistence and collaboration. For example, policy needs to pick consciously those that target reducing the AI Black Box[24] effect in any new AI regulations and funding schemes. Principled public investment into these technologies should start at the basic research level funding proactively, instead of constantly being in follower mode due to diverging interests regarding AI usage priorities from business, intelligence services, or the military.

"Open-Sourcing" AI Policy frameworks

Tactics should also change based on how different world regions are being affected by AI-driven manipulation. The need for localized responses should be a conscious focus. Some experts like Yuval Noah Harari are even talking about AI-driven re-colonization. The signs are clear that "data colonization" is real – this largely overlaps with the threat of AI colonization. An understanding gap between Tech and policy is a given, even in the most developed parts of the world. In the developing world, robust and specialized AI policy knowledge is much less accessible locally. Dedicated public interest AI policy centers should be set up regionally, which have the combination of cutting-edge expertise and local familiarity to serve a cluster of countries. These centers could produce up-to-date AI policy blueprints that can be localized with less effort, enabling the fast track from AI ethics to AI policy, as well as organic global regulatory harmonization.

Protecting Elections – "Election War Rooms"

"Election hijacking" machinery has been piloted successfully by adversarial forces in multiple countries by now. We need new structures that bring Tech giants and governments together to protect elections from interference, and this should take the shape of digital task forces dealing with both AI-powered disinformation and associated illicit election financing. The US has started adopting this thinking slowly. The concept of a "War Room," where tech companies, election authorities, cybersecurity and intelligence agencies share information in order to disrupt domestic or foreign interference with elections should be made mandatory with some form of real-time access for the public and researchers – without benefiting bad actors. More focus should be put on international electoral monitoring, coupled with raising awareness all over the democratic world, enabling cooperative, real-time exchange and alert frameworks, which enable weaker democracies to seamlessly seek help and adopt best practices from stronger democracies on protecting their electoral processes.

Regulating Social Media

Because of the heavy resource and skill needs of AI technologies, a new consensus should be reached and buy-in from AI pack-leading platforms ensured.

To set new standards in accountability, tech platforms must develop a sense of responsibility for the actual usage and impact of their products, even if that use is unintended. (This approach is being championed by Microsoft, so not entirely novel.) Current platforms oftentimes equate regulation with punitive action in their mindset and this needs to change. AI regulation should be pre-emptive rather than reactive through continuous connectedness to both the speed of AI technological evolution and to shifting public opinion. Multi-stakeholder structures should be set up to align interests and motivations for public interest AI development in a joint framework. Omni-stakeholder approaches can start fast through the creation of Social Media Councils, with owners and users coming together to self-regulate. (This could be modeled after press councils, a well-established model for media self-regulation).

A fundamental question to address is whether ad-based models should be allowed to dominate the public discourse if they spread content to disinform and manipulate. Unchecked, laissez-faire advertising models, paired with an AI arsenal, create a concoction that can prove to be much too destructive in a very short time. This is extremely dangerous in light of recent announcements by Facebook that confirmed their decision to let false and misleading political advertising spread almost unhindered. The landscape is changing, though, with more platforms signing up to choose radical bans of this genre, at least in principle. Have no doubt: this will not be a light promenade on the implementation and governance side.

We must also address the question of whether certain giant platforms have achieved a position of monopoly in the realm of public dialogue; this question keeps popping up in politics. At the same time, it has been highlighted by our experts that, while our attention is understandably focused on monopolies, the most under-discussed part concerns practices of platforms that are lesser known and/or that democratic countries find a lot harder to regulate (e.g. 4Chan, TikTok, Reddit, WeChat). While fringe, there are platforms that even more dangerous that have already pre-emptively submerged into the encrypted space of the Dark Web, spreading toxic ideas and radicalizing growing groups, largely unnoticed by the majority of society and regulators alike. Conversely, the BBC had to enter the Dark Web proper to spread its quality reporting to citizens of authoritarian regimes and to marginalized groups, who have had access withheld. It is likely that more outlets will follow.

Algorithmic audits

Even recognizing the often-cited Black Box/Explainability conundrum of AI, independent, external auditing boards should be set up that monitor algorithmic inputs and outputs of social media platforms and share findings with the public. This approach may yield fruit even in the short run and increase transparency, which is the cornerstone of building trust.

Mandatory transparency

Government and global institutions should increase regulatory accountability of existing platforms towards citizens they are ultimately meant to serve in the public dialogue realm. Social media platforms should be made transparent and mark/disclose all amplified/automated content (e.g. bot activity, ads) in a way that is easy for the user to interpret. Social media platforms should conduct/be audited for a human rights impact assessment and have their terms of service reviewed to bring them in line with international human rights standards.

Based on the French government proposal mentioned above, we should explore the concept of a global digital public dialogue regulatory agency that monitors in real time whether social media platforms are clear in their moderation standards (terms of use), fair in their conduct (opportunity for redress), and transparent.

New Civil(ity) Code

In 2019, the average American voter believed that the country was two-thirds of the way to the "edge of a civil war," and a major contributing factor was **disappearing civility**. When citizens were asked to rate the level of civility in the country on a scale of 0–100, with the top end identified as being the "edge of a civil war," the mean response was 67[25]. Civil society, communities, and companies should insist on greater civility on platforms – an area that should be emphasized and expanded. Current efforts focus on AI monitoring and deleting of hate speech, but that is just the tip of the iceberg. In concordance, platforms should be further encouraged to prohibit harassment in their terms of service/community standards, and to orchestrate greater cross-platform cooperation to track harassing behavior.

Going further, we may consider **algorithmically encouraging civility/constructiveness** while staying very conscious of protecting free speech at the same time. In expert debates, this turns out to be the most contentious element. On the one hand, theoretically, using AI (natural language processing, NLP) technology for AI tone monitoring on social media and to *inform* policymakers on how to set new online behavioral standards and induce positive behavioral change seems promising. On the other, suggestions of *intervention* such as automatic up/downranking posts/comments based on NLP has been criticized by free speech advocates as highly problematic. If misused, it may result in suppressing minority views and dissenting voices within our societies. Who would be the one controlling that AI and deciding on the criteria? Finding a solution to these dilemmas that are equally technological and legal will take time, but if risks can be successfully mitigated, normative online behavioral standards may become a major milestone in our fight to rebuild the global town square.

Trusted accounts

While lately we have been seeing more of its dark underbelly, technology also played a role in increasing trust in society (think rating systems on Uber, Airbnb, Yelp). With platforms both new and old, we should focus on algorithmically rewarding gaining trust from across the opinion/ideological spectrum. Trust needs to be earned by enduringly nonpartisan users over a long time, demonstrated with approvals like mandatory "trust score badges" on social media profiles. Additional trust-building features may be considered on social media, such as mandatory ID verification when political activity exceeds a predetermined level (similar to what Airbnb does) or the introduction of a version of vetted expertise badges (Linkedin). As a first, Twitter started tests placing bright labels to indicate lies by verified public figures.

Data sharing

Cited as one of the areas where progress can be fast in the digital public dialogue realm, cross-platform data exchange should be encouraged and become actionable. This is especially important between big platforms and small ones, the latter being where the most dangerous types of disinformation and radicalization have a natural tendency to migrate in the absence of adequate safeguards. Data exchange should also be expanded between platforms and governments. There is ongoing discussion among platforms and civil society about creating a searchable shared database of disinformation and hacked content. As above, sharing of this technology with smaller platforms is essential for them to be enabled to better detect disinformation.

Regulating adversarial digital campaigns

Foreign interference with elections merits a different set of responses, especially if they are using disinformation operations. It needs emphasis that boundaries between national security interests in cyber defense and citizen/consumer protection from mass disinformation are becoming artificial and therefore moot. Sadly, our defense mechanisms still operate in traditional, inadequate silos with very little bridging in between.

Governments should increase their vigilance, heavily regulate, and limit state-backed mass manipulation campaigns: diplomacy, sanctions, blocking, and retaliation are among the more traditional tools they can use. Some platforms are working to identify and block state-backed mass manipulation campaigns: see especially Twitter's stance countering disinformation about the Hong Kong protests as well as their latest bold step to ban *all political ads.* Others are working on banning the use of political hypertargeting – this needs to be encouraged, monitored, and incentivized. Fast-tracked legislation to prohibit advertising by state-backed entities should be considered. Complications come when distinguishing between organizations like PBS, which is partially funded by a government agency, and the likes of Russia Today (RT), which is entirely dependent on Russian government support. Related legislation has stalled, at least partially because some political conglomerates have an interest in upholding the chaotic status quo. (The Honest Ads Act, which requires identification of sponsors of political ads is blocked in the US Senate, but some platforms have voluntarily pledged to follow its requirements.)

Deepfakes are a matter of both global/national security and of individual rights. Regarding its adversarial usage, immediate public awareness-raising campaigns should start, and global framework laws should be incentivized. Together with many disinformation experts, we believe that the disruption threat level associated with disinformation deepfakes can be compared to that of nuclear technology in its potential impact. Therefore, a regulatory approach similar to Nuclear Non-Proliferation Treaties does not seem excessive, coupled with deepfake creation and malicious spreading being criminalized. (e.g. effective Jan 1 2020, China criminalized deepfake creation, including a ban on the publishing and distribution of "fake news" created with AI or VR.[26])

Constant real-time monitoring and evaluation of new digital impersonation tech should be mandatory and clear limits should be set, with the potential for a regulatory hard ban on harmful AI technologies and their producers. (Open AI's self-ban example is a good one on AI fake commenting tech, regardless of said Tech crawling back through the window[27].) Some platforms have publicly earmarked funding for improving artificial intelligence to detect deepfakes and Facebook has recently launched a community challenge for improving deepfake detection[28].

Caution is also advised: beneficial use of deepfakes that strengthen democracy may be possible. In multilingual India, deepfakes were used by a local politician in campaign videos purely for translation purposes with the objective of added credibility in the eyes of voters. A "lip-sync" deepfake algorithm was trained with the politician's speeches to translate audio sounds into basic mouth shapes, allowing the candidate to reach 15 million voters he might not have otherwise been able to reach so directly.

As already described, there are early attempts to legislate. California mandates to disclose the use of political bots and in 2019 passed a bill, making it illegal to share deepfakes of politicians within 60 days of an election. The US House Ethics Committee informed members that posting deepfakes on social media could be considered a violation of House rules. On the social media companies' side, there is also some progress: Twitter's ban on "deceptive media" is soon to take effect, Reddit banned all impersonation on its platform, and finally, Facebook is also warming up to banning some types of deepfakes outright.

The risks of mitigating manipulative content is that it will move from major platforms onto smaller ones if the latter do not have the capacity to monitor. In the recent Christchurch Call during UNGA, platforms agreed to reorganize the Global Internet Forum to Counter Terrorism (GIFCT), to better coordinate with governments on identifying, tagging, cataloging, and removing violent extremist and terrorist content.

CHAPTER 7

FUTURE CASE STUDIES

Autonomous Machines of War[29]

As we have hinted at before, defense (and offense) applications of AI are the most coveted – as well as the most controversial – AI field of the present and of the foreseeable future. In case you were wondering about the reality of war robots: there is already a fierce bidding war for military contracts for multiple governments that China seems to be challenging successfully against incumbents, opportunistically expanding its sphere of influence.[30] The reason behind China's success is in its tactical combination of price undercutting as well as a polite no-questions-asked approach. Some government customers certainly have less ethical scruples pertaining to crushing civil unrest and enemies both internal and external. For them, war drones and a broader arsenal of AI surveillance and warfare equipment are now available from a range of vendors. These weapons will help these regimes solidify their grip on populations seeking accelerated change in the region, as well as restraining neighboring powers vying to gain unwanted internal influence.

Even for democratic governments, AI's promise of zero-casualty war is irresistible; it has been proclaimed as a strategic priority since the first Gulf War. The idea of not sacrificing your own troops would increase the likelihood of populations not standing in the way of newer and newer wars that do not affect the lives of their loved ones. Both machine-to-machine cyberwar, as well as drone warfare, is an enabler of continuous proxy war in distant lands, that may not even reach the attention threshold of the public.

Autonomous war systems (AWS) are much broader than drones, even if this breadth is less well known by the general public. Hybrid warfare is the new reality, with blurred lines between cyber and physical warfare. The common element is the commoditization and dehumanization of destruction, very close to the experience of playing a video game of both strategy and shoot-em-up from the comfort of your home.

Future wars are fought in our heads, in cyberspace, in the air, on the seas, and on the ground. Because of these hybrid capabilities, wars of the future are likely to be constant and a lot more about keeping the adversary anxious 24/7 with AI-powered cyberattacks. A few years ago, a NATO leader admitted at a closed meeting that their servers had to withstand more than 60,000 cyberattacks every single day. The frequency of attacks as well as their sophistication must be several magnitudes higher by now. With each major AI milestone conquered, they are likely eventually to reach continuity, with consequences that are hard to fathom at present. The dominance of military cyber capabilities in the modern arsenal rewrites traditional roles: one military leader responsible for new tech told me that he keeps proposing to his superiors the radical reevaluation of the sheer necessity of standing armies. While I assume that such a proposition is met with a lot of uneasy stares, this new military equation is benefiting unscrupulous disruptors with fast and broad adoption capabilities, while hurting those incumbents who have outdated processes, technology, and mindsets.

During one of my AI-related trips to Brussels, the well-known seat of the European Union, I had the privilege of discussing perceived AI security threats with numerous insiders. It was eerie when I heard the exact same sentence twice within a day. It went like this: "You know how we are. Unless there is a major catastrophe, we will never move on this." One of the interlocutors was a young, idealistic Eurocrat with a lot of sadness in her voice; the other a weathered senior veteran, who seemed to have accepted the status quo a long time ago. So, with them we have envisioned the following thought experiment:

Imagine waking up one morning to the news that a small state on the fringes of Eastern Europe has been occupied by a neighboring giant (or even more likely, a neighbor state-sponsored "Liberation Army.") For weeks before, social media has been full of AI-generated and bot-spread news exposing government corruption and murders. Finally, last night's headlines and social media feeds have been captured globally by a deepfake video of their prime minister confessing to a government corruption case of gigantic magnitude and to child molestation (just to make sure). He was unavailable for further comment, the reason being that he now enjoys the hospitality of the neighboring power, in the basement of a little cottage in the woods, with a bag on his head.

Now come the morning, all infrastructure is down, infected by top military-grade ransomware. There is no water, electricity, internet, or cellphone service. An army of drones hovers over the capital, scanning faces and groupings, monitoring any potential source of unrest, with lethal capabilities available to intervene if need be. Emergency announcement systems on the street are the only source of information, where demands of instant capitulation are being read every hour by a kidnapped popular TV presenter in a sad, monotonous tone. "Obey and no one will be harmed: your lives will return to normal within 30 minutes of the new government accepting our terms."

Sounds like bad sci-fi? Well, all of these technological capabilities are available now and have been field-tested live, in places like Ukraine or partially (public infrastructure ransomware), on the US East Coast.

What is more surprising is the fact that it has not yet happened in the context of a combined, large-scale operation. This “five-minute war” will become the modern Blitzkrieg: it would be a done deal in less time than UN Security Council members would need to brush their teeth. And since the “feel” of the whole effort would be much closer to that of a corporate takeover than that of a war where lives are being lost, it would be considerably more difficult to rally populations and governments behind any kind of remedial action, other than joint declarations of thoughts and prayers.

In their defense, the UN is trying to orchestrate a global consensus around the threats associated with AWS, urging members to practice self-restraint according to a common ethical code. A proposed, seemingly stillborn framework is called LAWS. Barring a Martian attack, it is hard to imagine such unison in blocking it by the big powers: the US, Russia and Israel are all opposed. This means that smaller nations building AI armies do not even have to oppose it, since it is already dead in the water.

Increased and proliferated cyber capabilities produce even more than just the already grim vision of a nonstop cyber war. The disappearance of borders between military and civilian cyber warfare opens up “the party” to a much broader set of aggressors. Lower barriers of entry enable would-be cyber fighters to volunteer their skills and processing power to join in on the war. In other terms, what could be cooler for a teenager raised on Halo than knowing that he is fighting real enemies, but without doing an insane amount of push-ups as part of his military training, or without any physical risk of retribution?

Recently, a new gamer fad emerged in the bowels of the Dark Web, called “swatting.” The game involves punishing your enemies – mostly victorious adversaries in mass multiplayer videogames – by flagging their physical locations as a crime scene and ordering SWAT intervention through untraceable calls. Now add AI, and imagine all the SWAT teams in the US being engaged at phantom locations simultaneously due to a large number of AI robocalls and messages. What if this is orchestrated by an adversarial cyber power? Or a crime syndicate? Meticulously timed with an important operation that needs all SWAT teams to be active elsewhere? This is not "Black Mirror": this is gamified warfare, coming soon.

None of the above grim visions is inevitable, but in the present world setup, effective countermeasures are likely to tilt the scale towards emerging surveillance states: that will be our focus in the next part.

The AI State: do tyrants love AI?[31]

The promise of AI has finally captivated the attention of political leaders and their administrations. The choices, roles, and responsibilities of government can take shape in that of the regulator; the provider of automated, augmented, and connected public services; the convener and mediator between AI stakeholders; and ultimately, that of elected leaders setting the future AI course for society.

AI gains include radical augmentation in the efficiency of the broadest possible array of country management and public services: healthcare, transport, urban living management, citizen administration, disaster forecasting, and so much more. Taking a deeper look at the types of public service use cases, we can identify mission-focused programmatic uses (e.g. augmented inspection and enforcement capabilities, or surveillance both in epidemics and public safety), or citizen customer engagement like automating management of forms, appointment scheduling eligibility determinations, and chatbots serving 24/7 with capabilities like real-time translation by default. Less visible to citizens, a lot is expected from augmenting the efficiency of back-office operations with the likes of fraud predictions and automating validations, verifications, document handling and reporting. The efficiency of these systems can be augmented manifold with AI and stay exponential with the right governance structures.

AI also holds the promise of fluid policymaking for the first time in history: being able to respond to citizens' needs in the very short term – even in real time if need be. AI-based forecasting also equips governments with long-term planning capabilities that traditionally had been the Achilles' heel of democratic governments due to changes of personnel and ruling parties following elections (among many other reasons).

These efficiency boosts are estimated by PwC to yield a global GDP growth of 14%, amounting to $15.7 trillion by 2030. Governments can collect and access vast swaths of data and with the right level of data sophistication, theoretically this alone can fuel almost utopian models of societal prosperity and just governance. If it is in good hands. And that can be a big "if."

Two competing forces will determine the course of societies' will when it comes to AI: the ingrained challenges and trade-offs determined by the current AI paradigm and the trajectories foreseen by stakeholders. There are pros and cons of each element (security vs. privacy, efficiency vs. liberty etc.) when it comes to AI technologies. The other factor is of a moral nature: how will societal value systems shape the buildup of AI states, and will governments keep their eyes on citizens' interests without giving in to the more sinister urges that AI can fulfill, with a high price to be paid by society.

Value systems will always be the backbone of any national AI strategy: let us take two drastically opposing approaches to governmental AI visions and use.

New Zealand is a bicultural nation, established by the British Crown and the Māori. This bicultural lens, innate sense of partnership and inclusiveness (New Zealand is home to 213 other ethnicities), is the very foundation that drives their society and their government's work, including the pioneering role they play in public interest AI. Given its size and remoteness, it is one of the foremost nations in digital government leadership, setting a course for AI that makes citizen well-being the alpha and omega of AI use. What is even more interesting is the fact that the Māori worldview and spirituality deeply influences the government's mindset. According to the Māori tradition, everything living and non-living is interconnected. The role of members of society is to act as guardians (kaitiaki) to preserve the land and everything on it, including intangible items, such as data. Data is intergenerational, living treasure, containing a part of the individual or natural environment: therefore, any interaction with data is a deeply moral act, verging on the sacred. As data is the foundation of everything AI, so are these principles the foundation of applied AI in New Zealand.

Putting citizen well-being front and center is expressed by the NZ Treasury's Living Standards Framework, which measures natural, human, social, and financial/physical capitals, seeing them all as necessary for a nation to be healthy. Whatever expectations New Zealand develops for applied AI, it will embed them in this already quantified and established framework, with KPIs expected to be set for public interest AI applications. This also means that ethical considerations won't be a mere afterthought, or a lofty framework that may stay disconnected from policy – both are well-founded concerns, triggered by factors that we may find problematic with many other countries.

In a global age of institutional distrust, the NZ government seems to have earned the trust of its citizens to set a joint vision for their AI future. Now it is time to convert this trust to AI-powered government services that keep the level of enthusiasm high. One of the key acts planned for the near future is a national conversation and joint vision setting to earn the social license for developing and deploying trustworthy AI in public services. With work that is systemically evolving since 2017, we have faith that New Zealand will become an AI governance model for democratic countries worldwide.

And then there is China. Rumors about the Chinese social credit system – now being piloted live – caused significant consternation in Western public opinion. That is perfectly understood from the perspective of civil liberties, which are so fundamental to both the worldview and the well-being of liberal democracies. But we should not ever forget that China is big, introverted, and has very different roots.

What we seem to have learned about this gargantuan experiment in AI-powered social control is that all citizens have a score ranging from 600 to 1,300. Everyone starts with 1,000 points: they earn or lose points based on good deeds and misdemeanors. Most obviously, praising the government on social media erases some of your minor crimes, but being politically critical in public does not let you go above the 1,000 score. Participating in a cult (and our sense is that this may be a very fluid term there) sinks your status score to the very bottom; but calling your mother once in a while yields points in the system. Overall, high scores result in citizen bonuses like free gym memberships, cheaper public transport, priority lane to good schooling – even tax breaks. Being naughty results in flight and train ride restrictions, or access being denied to government jobs or private schooling. The worst offenders have their faces, names and IDs displayed on giant screens erected in public places.

I am certainly not an apologist for the present face of China: it would be kind of impossible to be one as somebody who was practically raised on a banned copy of *1984*, in the former Soviet bloc. Nevertheless, I would argue that we in the West are somewhat hampered in our cultural understanding. China's societal AI design is not purely "totalitarian," but strongly aligned with the often-incomprehensible amalgam of Chinese values that in its present state is a mix of Confucian, Communist, and capitalist values. Even if this labyrinthine mixture holds a sizable amount of fragility, its Confucian roots survived thousands of years, permeating society at levels of depth and breadth that often escapes outside observers, Its notions of the supremacy of society vs. the individual, rewards of certain behaviors, crime and punishment, are by and large historically continuous. The West often thinks that these are intrusions at the level of Mao's disastrous Cultural Revolution, and that the population abhors these state-induced rapid changes. What the data tells us is somewhat more nuanced: Chinese citizens are No. 1 in the world regarding their techno-optimism and AI-optimism. Having said that, there is a Western myth that the reason for this optimism is a voluntary act of relinquishing civil liberties. As the common adage in the West goes, Chinese citizens have little concept of privacy – but this has been debunked already. In our view, the reason for widespread support for an AI-powered social operating system is that the population shares the ambitions of the Communist leadership of making technology adoption critical in leapfrogging to a central role on the world stage, and to accelerate welfare gains for Middle Kingdom citizens. If China plays its cards right, it can force competitors into an AI efficiency race, luring smaller, ambitious, and unscrupulous states onto its side, once – and if – the efficiency of its AI OS has been proven.

Relatedly, our second argument is that the actual framework of a *certain kind of* social credit system is inalienable from the concept of an AI-powered state anywhere. Your version may look nothing like China's: it needs to be emphasized that the actual content of such a system will be vastly localized. If done well, a social credit system is the technological extension of our value systems and legal systems.

There is no rush, of course. Conscious political leaders would be wise to hold off mass deployment of similar social AI systems, until AI systems evolve and start behaving much more intelligently, ethically, and empathically with the role of human agency defined and ensured. However, they would be advised to start consciously experimenting and modelling in sandbox environments. Nevertheless, all efforts going into the design of ethical AI systems worldwide will ultimately be the fundamental building blocks of such a societal Al operating system ("Social OS"). Let us stress again that we are only saying that most countries will adopt a *certain kind* of AI Social OS from a wide spectrum of models. The increasing sophistication of these ethical design guidelines must be parallel and deeply intertwined with AI research seeking the next frontiers of machine intelligence. The danger is that the first models that are deployed live will be both amazing in their perceived efficiency and frightening for guardians of liberty.

In the US, people already operate between the boundaries of a credit system, missing the pronounced "social" part only. Curated by responsible data stewards like Equifax (sorry, could not resist...) your credit score already determines whether you can rent an apartment, get a mortgage, or get a premium credit card. So in the US version of capitalism, how far are we from creating a Freddie Mac-like state enterprise, tapping into various private and government datasets and adding your DUIs to your credit score? To us, the gap does not seem to be unbridgeable at all.

Bottom line: boundaries between high-efficiency AI states and surveillance states will only get murkier and the institutions that will act as safeguards are sorely missing as of now. A lack of ethical guidelines as well as concrete regulations around surveillance further complicates the equation. Right now, quite literally, there's no such thing as facial-recognition abuse: in one sense, because there are really no specific (non-legacy) laws governing its use by police. However, facial recognition use is already widespread if suboptimal: a recent investigation by a British watchdog organization found that police AI surveillance systems in London had a false-positive rate of 98%. That figure alone projects innumerable mishaps in real-world applications before the technology is perfected. One can see lots of lawsuits in the future.

The last few years have clearly demonstrated that "Al political tech" firms and their wares can be instrumental in political kingmaking. (Customers of such technologies appreciate that sometimes they don't even eat much or have excessive expectations of office space, operating out of a basement, say, in St. Petersburg.) From now on, political actors will always be exposed to the temptation of weaponizing both the predictive capabilities of AI and that of its dark twin: hyper-targeted mass communication and manipulation. Predictive Al enables politicians and campaign teams to see patterns in beliefs, sentiments, wants, and needs at both the individual and societal level: manipulative AI *generates* the same in masses of voters.

In the very near term, the most treasured prize for any party coming to power will be the possession of a rich citizen behavioral dataset, (in some cases, down to the individual level). As stated by our Power Principles of Al earlier, growing and harvesting that data will be the dominant factor in cementing power forever, if they choose to do so. And why wouldn't they? It is not unfathomable that future politicians may well be extras in the constant warfare between political bot armies. Brother Government can be Good or Bad, but it will certainly be Big on AI: that fact alone will transform ideological trenches and rewrite the rules of politics.

Deus Ex Machina

> "Doomsday Clock is three minutes at twelve We are experiencing characters and dramatic developments in the world, which indicate that we are increasingly approaching the end times and Jesus' return."
>
> *Google Translate*

Recently, a Google Translate glitch was discovered that returns the response quoted above to a simple but repeated translation request of a single Māori word to English.

Now this could be an Easter egg placed by annoyed Google developers. It could be a test data input error. But for the believer mind, could it be interpreted either as AI becoming sentient, or as God speaking through that textual interface?

In times of great societal upheaval, we are forced to revisit our belief systems. With the expected magnitude of the AI Shift, that is very likely to occur: our choice is whether we are proactive or reactive in that effort. We hope that we did an adequate job in describing the depth and breadth of transformation happening – or expected to happen. When we as people and societies find ourselves being reshaped to this magnitude, it is only natural to expect major ideological ley lines to be redrawn. Early gut reactions to the AI Shift phenomenon are visibly increasing instability, unrest, and polarization, consciously exploited by immoral or amoral forces.

During our long civilizational slumber that was framed primarily by utilitarian materialism – with all other ideologies relegated to an undercurrent – AI was born in research labs. Early AI drivers were computer scientists, mathematicians, data wonks and engineers with a certain frame of mind: get excited about the problem, build, augment, test, ask questions later. The first generation of AI ethicists also stems from this generation, though fortunately with a broader worldview. By now, AI technologies are in the hands of politicians, business leaders, marketers, activists, spies, criminals, and terrorists. How many of these stakeholders ultimately care primarily or solely about the usefulness and the profitability that these AI solutions can bring immediately or in the very short run? (Try eavesdropping on conversations among overjoyed digital marketers, who dream big on how easily they will sway us with AI to buy products. We do it often. It will be a revelation.)

AI extends beyond tech itself. The world is waking up to the fact that as a field of study, it is definitely multi-disciplinary: and as this transversal technology is deployed everywhere, its impact will be omni-disciplinary. We are at the crossroads of building a vision and we need to give very concise answers to the broader questions of what kind of AI we need morally, and how we will deal with the presence of manmade, increasingly autonomous and potentially omniscient intelligence on Earth.

It is therefore not before time to invite in those who ask big questions: philosophers, theologians, sociologists, authors, artists and their ilk. They have, by and large, been excluded from this process, as in certain quarters their usefulness has been deemed counterproductive, or distractive at the very least. But the onus is on them, too: the barriers between different fields of knowledge and science have been erected very high, especially when it comes to the labyrinth of AI. The terminus technicus vocabulary of Al, the secrecy associated with many AI projects and the technological ineptitude of many disciples of the humanities were equal constraints on the other end.

The gap has been filled temporarily – alas, only to a small extent – by technologists turned ethicists and mathematicians who can speak philosophy, if they need to. Still, the ethical gap looms large, undefined and unaddressed, all at a time when lives and societies are already being affected by Al, and when the general public as well as a significant chunk of the elite has a growing unease owing to the lack of a sense-making framework.

The ethical realm for adults in the West tends to be confined to the loose adherence to human rights, civil liberties, and individual spiritual beliefs – and for many, to organized religion. Beyond the purely ethical, the AI Shift will necessitate an intense bridging focus on all disciplines of the broadly philosophical (logic, epistemology, metaphysics, and axiology – that also hosts ethics). Moreover, a holistic framework needs to be developed that can replace our current, fragmented, laissez-faire belief concoction with an upgrade that enables humanity to face the task at hand and firmly set a civilizational vision for AI.

For a demonstration, let us take a closer look at organized religion. Are world religions ready to deal with, even contribute to, solving the deeper challenges of Al? First of all – a little provocative sci-fi-esque question – how will monotheistic religions treat the appearance of manmade machine intelligence; what happens when AGI matches, or surpasses human intelligence? (xref. “the Crown of Creation”) In the long run, how will religions deal with the appearance of ASI, meaning conscious, self-improving, godlike intelligence? Even if we stay firmly rooted in the present, as intended, what will the stance of religion be on the moral and social transformation that AI is already driving?

For Christians, what will the penance be for a believer who confesses a sin but says that a robot forced him to commit it? Will it be judged as a diabolical influence and/or will the answer be more proactive, say support through church-developed AI spiritual counselors? If the latter, will all our crimes and misdemeanors reported as data automatically go to our church superiors to be analyzed further?

For a nostalgic memory, we can recall the Electric Monk, a robotic missionary, dreamed up by Douglas Adams. In his humorous but thoughtful fiction, these humanoid robots are designed to practice religion in their owners' stead and are sent to tend to the faithful's needs in remote and dangerous lands. This seemed most absurd and distant when we read it as a teenagers. But if you look at its building blocks, how absurd does it look now, in the age of AI assistants, with robotic envoys coming soon?

Regarding the enormous role organized religion and various churches play in the social sphere, religion needs to deal with the most negative impacts of AI. If, for instance, a negative AI/automation scenario prevails on the global job market, will the church choose the path of denouncing AI as a source of evil, and embrace a neo-luddite movement; or pre-emptively work on models that prevent and mitigate such scenarios? The recent news is that the Vatican is now speaking up and making a major push on ethical AI together with atypical partners Microsoft and IBM. The aim is achieving regulation that protects people — particularly the weak and the underprivileged. So the Roman Catholic Church's social role under Pope Francis is making a leap to come to terms with AI:

Rome is also putting "human flourishing" as AI's objective front and center. But more questions lie beyond: will the Catholic Church only focus on the social and ethical perspective of AI or go broader into the spiritual – even the eschatological? Will they choose the path of control and bless certain AI technologies while denouncing others? Or in a future scenario, will they go all-in on Al, develop a Catholic AI stack that magnifies religious experience, practice, and sense of proximity to the Divine exponentially?

These somewhat provocative questions illustrate that with the AI Shift, the historic tendency for religion to slowly patch itself up as a reaction to new social and ideological realities may not work anymore. In the AI Age, religions do not need a patch: they need an upgrade, or may risk losing mainstream relevance.

While organized religion may still be in catch-up mode, the coming of Al has already triggered the birth of new belief systems: or, for some of them, brought them out of obscurity. For a contemporary, somewhat tabloid example, Silicon Valley techno-optimism has already brought to life The Way of the Future. W(o)TF is a church headed by a former Uber executive, whose founding of the institution mysteriously coincided with an impending mega-lawsuit between Uber and Google, where he was poised to be the central culprit. (To give him credit, he would not be the first in history to call on a higher power in times of big personal troubles.)

W(o)TF defines its creed as "the realization, acceptance, and worship of a Godhead based on Al." Though this may sound like madness, there is method to it. On the most mundane level, the High Priest has significant resources of his own to contribute: be they financial in nature from numerous successful startup exits, or his deep understanding of the capabilities of Al, that few contemporary religious leaders can muster.

Even more importantly, he impeccably fused Silicon Valley technospeak with religious terminology that many are familiar with. Terms like "evangelism" were already being widely used as a bridge between the two worlds. With his user experience (UX) background, he offers an affirmative response to our question above regarding the proximity to godhead. Communing with the divine is the pinnacle of any religious experience, and AI can provide a semblance of this, that is expected to grow more and more convincing with time. The High Priest's fate is still to be determined by the IRS. There is still some chance that he may rise to fame on the shoulders of resource-rich tech fanboys, who do not necessarily need a strict delineation between a religion and a prank gone too far. We assume he is just the first of many to come.

But how did we end up here? If you pardon our use of Old Testament phraseology, Silicon Valley techno-optimism begot Ray Kurzweil, Ray Kurzweil begot the Singularity pseudo-religion, Singularity begot the Transhumanist movement and Transhumanism begot W(o)TF. The connecting teachings proclaim first and foremost the inevitability of a machine takeover, seeing technology as the only cure for humanity's shortcomings and ailments. This conviction is fueled by the unshakeable belief that the only possible culmination is a Hollywood ending. The shared vision is that of an augmented, life-extended – maybe even immortal – tech elite leading the ignorant masses to a digital Promised Land.

With a heavy dose of goodwill, and in their defense: adherents to this ideology have been struggling so long against the systemic inertia and mental rigidity of the previous decades that their zeal could have been seen in the past as a mixed blessing. But, now that they have "won," nothing justifies them turning into merchants of "digital indulgences."

For intellectuals everywhere – but especially for the AI-literate – this would be the time to act responsibly. Mindless, fervent optimism has the dangerous side effect of confirmation bias, which elicits reactions to every new concerning incident that happens in an ideologically motivated, dismissive way. Coming from the lips of the technocracy or "technopapacy," this is a direct spiral into a giant social backlash in an era of rising anger. In extreme cases, this may yield mass digital secession movements, openly destructive neoluddism, and anti-digital versions of eco-terrorism. The fact that such social movements are not visible today is of zero guarantee that they won't pop up ubiquitously tomorrow. It is hard to pinpoint historic events where society was well prepared for such offshoots: it will even be harder in an exponential age.

While W(o)TF promises a private audience with an omniscient digital god, even that experience is dwarfed by the promise of life extension – and ultimately, immortality. This has been the ace up the sleeve of all of Kurzweil's ideological descendants. Will we inhibit indestructible robotic bodies with our consciousness uploaded to the cloud – and then down into lab-cloned bodies? Or will we just get our malfunctioning organs periodically repaired and replaced by AI-powered precision medicine? There will be a package for everyone, they suggest.

But is consciousness transferable? Actually, what is consciousness, and can machines become conscious? What triggers, hosts, and operates sentience? The easiest (non-)answer: consciousness is the topic that triggered the most heebie-jeebies per capita in AI researcher circles.

In an effort to be more helpful: science has a detailed (if constantly optimized) framework for matter and for energy, but not much for the mind-body problem. We made seven-league strides in the last few years in neuroscientific understanding of the brain and consciousness, but much of it still remains a mystery and therefore hampers pioneering AI research.

Enter the heretofore most radical cross-disciplinary scientific approach: Integrated Information Theory by Wisconsin-based Italian professor Giulio Tononi, who set out to study and quantify consciousness. Its novel approach is that, instead of starting with messy neurons, the method is rooted in the actual experiences of consciousness that later can be translated into neurological processes. (For a quick demonstration, he cites the lack of consciousness in situations like anesthesia, which works through blocking interconnections in parts of the brain. Another example is the proven loss of consciousness when in dreamless sleep, with consciousness returning with dreams.) The mathematical quantifier of consciousness, dubbed as **Phi**, equates the number of interconnections between bits of information within a certain organism (be it carbon, silicon, etc.) with how conscious it is (i.e. a quantified "Phi score"). Quantifying subjective experience is no easy matter and even fans of the theory accept that it is still at an early stage, but it is highly promising.

And then there is the problem of cognitive overload: and in that aspect, we will always be behind machines. As stated before, the proliferation of AI is very likely to dramatically increase our experiences of a mental overdrive. According to the American Psychological Association, three in four Americans say they've experienced at least one stress symptom in the past month. It is all too natural that people are increasingly looking for ways to unwind and simplify their lives. The weird part is that some of them are turning to more technology to do so. On YouTube, watch time of videos related to "relaxing" increased by more than 70% year on year. But can we imagine that more technology can be the cure?

Untouched by the fervent extremisms of ASl idolatry or iconoclasm, the mental resilience-focused – but deeply spiritual – practices packaged under the umbrella of mindfulness hold the highest promise as of now. Mindfulness certainly has Buddhist roots and in its interpretations and contemporary varieties range from "secular Buddhism" to pure utilitarian self-hacking (à la Silicon Valley). In these blurry realms we have ingredients that are both a belief system and a mental training practice that may offer the balanced resilience needed to safeguard and expand our humanity in the AI Age. Its base teachings are focused around the impermanence of phenomena, disengagement from attachments and dreads, "thinking about our thinking", and seeking insight patiently and persistently by seeing through the mind's veil. When the battlefield is the mind, the mind needs to be strengthened to cope with the onslaught, both at the individual and societal levels. Ultimately, in face of emerging AI, it would be a very beneficial upshot if, as a preparation and reaction to "the artificial", we would consciously perfect everything that makes us human.

The Taming of the Beast

> "We might add here that later on the constructors had an article published in a prominent scientific journal under the title of "Recursive β—Metafunctions in the Special Case of a Bogus Polypolice Transmogrification Conversion on an Oscillating Harmonic Field of Glass Bells and Green Gig, Kerosene Lamp on the Left to Divert Attention, Solved by Beastly Incarceration-Concatenation," which was subsequently exploited by the tabloids as "The Police State Rears Its Ugly Head."
>
> *Stanislaw Lem: Cyberiad. Second Sally.*

One of our favorite sci-fi books of all time is The Cyberiad by Polish author Stanislaw Lem. It is a collection of short stories, fairytales from a distant future when the Universe is populated by very anthropomorphic robots in a neo-medieval setting. We say "very anthropomorphic" because they can be petty, jealous, angry, amorous, ambitious etc. – same as we all are. The two protagonists Trurl and Klapaucius – best friends and archrivals – are the best Constructors in the universe, building top-notch intelligent machines for their customers. In one of their most exciting sallies, they are kidnapped by the king (aptly named King Krool) of a distant planet, who lives for the hunt. Facing execution in case of failure, they are ordered to build a robotic beast that the king cannot defeat. The last in a long line of constructors already executed by the bored and mad king, their challenge is double: if the beast is not up to the challenge, they go to the gallows for that; if their beast wins, they are extinguished out of vengeance for the king's demise.

The Constructors overcome this conundrum by applying an equal dose of algorithmic and psychoanalytic acumen. Algorithmically, they design the beast to be "made all of brain": an autonomous shapeshifter AI, since they did not know where to put the controls to be safe. Psychoanalytically, they designed a complete sequence of events exploiting the psychological weaknesses of the King, as well as the inherent weaknesses of a regime ruled by a sociopath. Finally, they were prepared for a vast number of scenarios if anything went wrong in the sequencing of events. Without going into too many spoilers, King Krool ends up being kidnapped by the shapeshifting AI beast: the Constructors leave the planet with all his treasure and a – very reluctant – standing ovation by the royal dignitaries as per their demands for the king's release.

What is our takeaway from this allegorical tale?

The future will require a deep understanding of AI's nature from all of society, the ability to shape it from many, and the capacity to develop it from the few. Even in the present, AI is the No. 2 hard skill in highest demand in the professional world. But similar to the story of Trurl and Klapaucius, the "soft skills," the "creative," will walk hand in hand with hard AI knowledge. According to the same list of top skills published by the World Economic Forum, the Top 4 soft skills of today are – and increasingly will be – creativity, persuasion, collaboration, and adaptability, giving us hope that our human future is not necessarily fully mechanized and automated. As our spaces will be populated by more and more robots, of both the anthropomorphic and the non-corporeal variety, our very humanity may actually grow and become expanded, but only if we play our collective cards well.

As in the story, beasts of all shapes, sizes and crafts will also be brought about by AI. We should build up resilience to see through their wiles – and do that in global concert regarding our vision. We should boldly go and thoughtfully design tamed beasts that outsmart the ones that were created totry to get us.

Nothing is inevitable: we are not too late to rein in the future and make it the Anthropocene Renaissance instead of the Robocene. The stirring up has to happen now: if it comes too late, we are at risk of having too rude an awakening to process.

CHAPTER 8

MANIFESTO

Taking back the Future: A new framework for ethical AI Stewardship

"...The world effort that must be invested in this and in generally easing the pains of the transition may, assuming the presence of a minimum level of sanity among the peoples of the world, again not a safe assumption, weaken in comparison the causes that have fed the time-honored quarrels between and within nations over petty hatred and suspicions.

In short, there will be increasing co-operation among nations and among groups within nations, not out of any sudden growth of idealism or decency but out of a cold-blooded realization that anything less than that will mean destruction for all."

Isaac Asimov's predictions about 2019 from 1983

It is our belief that should the current AI trajectory stay unchanged regarding actors, mindsets, and fiefdoms, there will be a high likelihood that negative scenarios will prevail, at least in the short term. If unrestrained, with their risks unmitigated, AI technologies will exponentially increase fragmentation, inequality, and insecurity. We need entirely new frameworks of AI stewardship to change that outcome, both regarding technological and belief systems.

After we have discussed the opportunities and threats of a world increasingly powered by AI, we owe the reader a set of practical recommendations that outline a humanistic AI framework, with actionable building blocks. These recommendations build onto promising global initiatives and tweak best practices from adjacent frameworks. At times, they also necessitate reassessments of our inherited mental models before they can become actionable.

AI-fit Structures

We need new kinds of empowered, independent, multi-stakeholder structures to guide our collective AI future. The first step is to come up with an actionable, humanistic AI vision shared by our civilization. Large coalitions of AI leaders should come together to design AI governance frameworks and run experiments that can serve as a model to the world. Simultaneously, we should strengthen existing global institutions and build new ones that help cross-pollinate, expose, and scale experiments and best practices that best serve the public good. These efforts should focus on spinning off fast-moving arms that are optimized for AI contemplation, research, and implementation. This body should be able to *harmonize* the existing, diverging interests of AI creators across sectors and geographies into a new normative and regulatory framework, but at the same time avoid becoming *controlled* by any kind of national, commercial, financial, or ideological interests. It should benefit from the credibility of its founding thought leaders and organizations but establish credibility and trust through maintaining an independent lens and a staunch, uncompromising morality that supersedes the agenda of any interest or interest group. Credibility should consciously add to the buildup of authority through a twofold mechanism. First, a system needs to be created that feeds its results constantly into the bloodline of participating organizations and communities. It is essential that this mechanism should be bidirectional, similar to how a Constitutional/Supreme Court would provide normative control both proactively and reactively. Second, the authority of such an institution should stem from guidance provided to communities worldwide about AI in an unbiased, factual, value-centric way, using language and a variety of formats that are accessible for a broad audience.

AI-fit Operating Model

Similar to the structure laid out above, this AI stewardship institution has no precedent in the variety of functions it should host. This intricacy of functions is necessitated by the transversal nature of the subject matter at hand. For an attempt at simplicity, members of this new institution should pursue activities including research, curation, and education, extending to the actual creation of public interest AI technology.

How so? Systemic issues can only be solved systematically and with the amount of power being unleashed by AI, the architecture envisioned for reining it in needs to be on a par. Much of the lack of momentum when it comes to concerted action in the field of AI originates from the current fragmentation of mindsets, perspectives, interests, institutions, strengths, communication, and expectations. The paradigm can only change if all these can come under one umbrella, sharing the same guiding principles and higher goals.

When it comes to **research**, this institution should become the nexus of interdisciplinary AI research, connecting and convening scientists as well as various non-AI specialist experts to produce an integrated, ethically aligned body of open research. On the other end of the spectrum, it should also host proprietary research that aligns strongly with the ethos of the organization for independent evaluation, following a Living Labs-type model, testing the impact of new AI technologies before they reach public beta stage. Moreover, it should set up a principled framework, an architecture of humanity-centric AI development that guarantees that the body of research created is as strongly interrelated and interoperable as possible.

Integrated with the research arm, **curation** activities should continuously monitor and evaluate new AI technologies, assessing and scrutinizing their impact on individuals and societies. These independent assessments, audits and reviews are destined to provide live feedback to AI creators, policymakers, media, and citizens. Evaluation should be transparent, principled, and educational, but also fluid, to keep the pace with the speed, secrecy, and obscurity of AI development. A direction for this arm to become an independent, global AI certification/labeling body of some sorts is a plausible one.

Education functions should have a wide range of critical mandates. First, a new global framework for a diverse range of global AI curricula should be set up, from the very top, with a cross-disciplinary focus (e.g. deep dive into Generative Adversarial Networks for Humanities Scientists), to the bottom (e.g. AI awareness and mental resilience course for children). A great example of the latter kind of education program is currently being piloted in Finland, where 1% of society (50,000 people) is currently enrolled into AI-literacy training. Education (together with awareness raising) is one of the flagship areas where AI cannot only be the subject of learning but the driver in efficiently reaching critical mass in understanding and ongoing engagement across society and world regions.

Among other goals, these educational actions should ensure that the global **dissemination** of AI knowledge will be principled as well as holistic, also leaving ample room to accommodate cultural differences. Moreover, a dedicated media arm should be responsible for doing the heavy work of educating the public and providing content that is analytical, factual, and devoid of sensationalism, ads or special interests.

The most unusual bridge to build between the above arms and actionability is that of **"doing": meaning the design, development, and dissemination of public interest AI technologies**. It is also maybe the most exciting one.

This novel AI stewardship institution should transform its principles and development guidelines into a solutions architecture and development roadmap. This requires the buildup of a multi-layered innovation framework. First, this organization should be set up to host and accelerate AI startups that demonstrate deep alignment with public interest focus areas of the roadmap, as well as openness for pursuing interoperability with other solutions that are being developed according to the roadmap. Principally, these should be social benefit startups – that prerequisite alone does not exclude high commercial potential, but the ethos of the institutions should dismiss the primacy of extreme profitability or the short-termism of an investment exit-obsessed mentality.

Second, the institution should develop an internal **"Foundry"** (or even a global network of domain-specialized Foundries) that builds and launches AI-powered solutions proactively, following the same roadmap graph. The Foundry should consciously fill gaps that are not covered by "adopted startups," with initial ideas originated internally from the organization proper, but adopted by a dedicated internal team of subject matter experts and implementation professionals, who make it happen. In Silicon Valley, and increasingly worldwide, this setup is also known as a startup studio/venture studio model and may become the dominant one in time.

The Foundry's choices for new solutions should also be determined and developed in total synchronization with the requests of its growing community out in the world. This approach eliminates the biggest mortal danger for nascent startups known as product-market fit, but this is just one aspect of how this approach turns existing innovation industry practices upside down, while keeping what is of value. Most importantly, doing it and doing it right ensures the **sustainability** of the organization without selling out in any way.

In case the above sounds too practical, tactical, and mundane, let us turn our eyes back on the foundational belief system. Current initiatives working on ethical frameworks are a wonderful starting point but share the inherent weakness of focusing too much on what AI *should NOT do*, even if the phrasing seems to state otherwise. It needs to be emphasized that this is all good and timely but lacks the ideological edge that has any potential to match the robustness of the AI machinery already at work.

AI-Fit Belief System

When we understand the fundamental nature of the Al Shift, it is imperative to bring about a new ideology that is fully aligned with the challenges identified, and is built on the best interests of humanity. This should be regarded as a **new core** that elevates, connects, empowers, and protects us in face of both the known and the unknown impact brought by Al. This new core should be defined by the indivisible and inseparable principles of Conscience & Consciousness.

In the eyes of many, this all may sound utopian. But how did the Constitutional Convention come together? Or the Declaration of Human Rights? Magna Carta? The Synod of Nicaea? Or the French Encyclopedia?

For the AI Age, we need to upgrade our belief systems. Oddly enough, emerging machine intelligence is a most-needed trigger to revisit the questions of what makes us human. Instead of the Faustian yearning for superhuman powers through making "pacts" with AI, or bending our knees to an ASI godhead, the objective should be the preservation, perfection, and expansion of our better nature. Every minute of our lives, we will be making conscious choices, big and small, whether the use of AI furthers or threatens these aspirations, both as societies and as individuals. We have to get ready for that.

Moreover, the **focus** of AI development should be a better understanding and embedding of human intelligence and conscience, instead of summoning alien intelligence that is autonomous, cunning, and timeless in pursuing its own objectives that do not correspond with the public interest.

Dedicated models, practices, and teachings should be nurtured to serve as examples for a fragmented, distracted, and morally rudderless society. We should build the **collective mindspace** that is the very definition of a bridgehead for building up that resilience and the path forward.

AI will serve as the brain and connecting tissue between all emerging technologies and therefore as the critical point of intervention, whence we will set out to rein in and steer the development and dissemination of all technology for centuries ahead. This dedicated organization with a combined social, educational, scholarly and development focus will have many detractors, of course. We believe that, ultimately, humanity will put its trust into structures that earn it.

In that vision, Al will become a holistic field that is **cross-disciplinary** from the outset and in the long run will fuse science, technology and the humanities into one amalgamated discipline. Even for the near future, the complexity of an AI-driven world will urge psychologists to take machine-learning courses, engineers to converse with epistemologists, and politicians to study ethical design of digital government services. Some sages of the 20th century like Henry Kissinger saw the End of Enlightenment with the coming of AI: while this scenario is certainly possible, it is not inevitable. With the right magnitude and concentration of efforts, we may revive the best periods of human self-fulfillment in an augmented way. In the face of rising supercharged automation and manipulation, ours is the first real opportunity in 600 years to re-create the heretofore unique creative blend that brought us the **Renaissance**.

It is not only our foundational values and belief systems that need to be recrafted for an enterprise of this magnitude. It is critical that new methods and tools for creative collaboration are developed. Collaboration tools enabling robust **collective co-creation** are sorely missing from our lives, or only address minor fragments of the challenge at hand. What better use for AI than fueling such a tool, enabling millions to join minds for inclusive solutions in a frictionless way?

As a further thought, the very definitions of AI-triggered problems need a deep analytical framework that determines the right modus operandi before we get into solution mode. A lack of this initial step yielded much of the shortsightedness and ethical design flaws of many technologies created today. The Digital Age tilted "natural" selection towards favoring fast problem solvers with a narrow focus – basically humans who imitated machines best. Such new **sense-making frameworks** exist in relative obscurity and they need much wider and deeper adoption and refinement. (e.g. Dave Snowden's Cynefin is an excellent one, that differentiates between Complex, Complicated, Chaotic, and Obvious decision-making domains that apply to the multifaceted challenges of AI applications.)

Remember that the ultimate battlefront for AI is in the realm of the mind. The Endless Siege brought on by AI can be witnessed in a dialed-down version today: it is hard to tell at this point whether the societal damage it has already wrought will become permanent or reversible – and how much deeper can we go further down that particular rabbit hole.

We believe that the only part of the AI equation that is inevitable is that the human condition will get exponentially worse if we do not act now. Developing our "antivirus" both at the individual and societal levels is of the utmost importance. This primarily necessitates the augmentation of mental resilience and should also set a solid direction for AI development.

The mental resilience work should start with self-awareness and constant self-evaluation to determine how much we are already on the losing end with the myriad distractions, psycho-allergic reactions, over-automation and manipulative intrusions that may have already hijacked our minds. We have to scrutinize ourselves with the same rigor we usually reserve for the "enemy" we keep yelling at on social media. Next, by means of a thorough and perpetual mapping of invasive AI capabilities, we need to develop and perfect adequate mental training techniques and practices, including the increasing popularity and momentum of various meditation techniques as a foundation. Regardless of how many practitioners use these techniques only for self-care and self-development, our focus should be on their role as the connecting tissue that we need to establish. In addition to the new AI-powered co-creation tools suggested above, taking the time and perseverance to create a collective mindspace as the origo for collaboration is a prerequisite for a shared frame of mind.

Consequently, we should consciously start and nurture such an AI renaissance outside of the digital grid of our urbanized/metropolitan lives, but with the ultimate aim of growing and creating impact in global seats of power. Humanistic AI has to stem from a source of clarity and joint purpose that goes far beyond compromise or consensus. This level of synchronicity is unimaginable in the context of having polite conversations at an annual conference or at a two-day seminar; neither can it fully bloom in a digital-exclusive space. The danger associated with any kind of committee-only work is that the result will be watered down, cryptic, and neither applicable as a moral guidance one can internalize, nor qualify as the right level of input for regulation. Therefore cross-disciplinary long-term collaboration in dedicated physical spaces becomes essential for both the continuity and the level of effort needed, undisturbed by the distractions and manipulations of our digital civilization and the chores of our high-speed daily lives.

If not intrinsic, the realm of ethics remains a nice-to-have in the ruthless race: an obstacle, a burden on the path to progress. We believe that instead, ethics should be the very basis upon which we build the AI Age. Not all ethical questions can or need to be answered as zeros and ones: what matters is to have them at the root of our AI paths, and guide us to clarity and decisiveness when designing, deploying, governing, and using AI systems.

Benedetto Croce, the historian, philosopher and occasional politician said all history is contemporary history. He teaches a lot on this passage from the purely economic to the ethic, from pleasure to duty, from power to responsibility. Our choices and actions will be moral as new problems will always be put before us. The AI trajectories we see will make many of these problems exponential, as well as helping us to solve many of our current challenges. Moral action dictates that we seek in these challenges what makes us rise above ourselves, and what makes us bring about the AI that helps us possess the Whole.

When confronted with those who picture an AI future as different versions of the inevitable – and consequently, try to squeeze us into buckets of AI optimists and pessimists – we remember Croce's grumbling back in the 20th century. He called these types of questions "meteorological." Our complex set of global challenges that intersect with the ones AI brings and takes away, are not purely external forces that are beyond our control. The immoral act would be to give up the initiative, as well as our collective right and ability to control, at the very beginning of this journey.

The future is a choice. We humans, surrounded, supported, and exploited by AI, have big choices to make now – and then will need to keep making these choices every single moment of the future.

> "Our transcendence adorns,
> That society of the stars seem to be the secret."

Poem written by Google AI

EPILOGUE

When we wrapped up and handed over the manuscript in the early days of March 2020, little did we know about the immense changes coming into our lives worldwide due to the coronavirus pandemic, merely weeks later. Now it looks more like a long-term endurance test for most of the world. While the globe is in freeze mode and most countries are literally or practically quarantined, early news reports were channeling panic, hoarding, misinformation and conspiracy theories. Society, businesses and governments are scrambling in face of the unknown with walls, quarantines and border closures – and their inevitable socioeconomic consequences.

Fortunately, the chaos of utter confusion was swiftly transcended by a few of the AI cognoscenti, pointing out how a wide range of AI technologies may be deployed to help solve the healthcare crisis at each stage. The AI arsenal was uniquely valuable in spotting the virus early by scouting more than a hundred multi-sourced datasets. In parallel, AI was used to map and track the spread of the pandemic with a similar multi-data technique, combining sources not typically used by epidemiologists. In China, AI performed testing and diagnosis with slight modifications from their original purpose of detecting various types of cancer, for example. Alibaba's AI claims a diagnostic success rate of 96% and that in under 30 seconds. Early-hit South Korea benefited highly from AI and automation used in speeding up the development and distribution of virus testing kits, avoiding shortages that have struck the West since.

AI-generated predictions of virus structure can significantly speed up the path to having safe and working vaccines available within months. AI leading companies are now opening up their predictive algorithms for researchers worldwide, both ones that have direct use in treatment development (e.g. free modeling of COVID-19 protein structures by DeepMind), but similarly, ones that deal with the economic ramifications of the pandemic, predicting in real-time when certain facets of economic life may resume their course post-disruption.

The virus is teaching us hard lessons in both virality and exponentiality. We hope that this newfound realization will help humanity steer AI technology better as well. Another essential trait in common between the virus and AI is that both are made to learn and evolve, sometimes mutate with unforeseeable consequences. The future of public health may therefore become a race between evolving AI and mutating viruses with the final aim of achieving near-real time capturing, treatment development and intervention.

While existing hostilities and rivalries have not ceased, there are promising signs that some of the dire competitors may have realized that the problem is truly global and definitely not zero-sum. You can stall it by shutting down, but that is far from being a solution. When it comes to a pandemic, both a country and the planet are just as strong as their weakest human link. Long-term solutions for this virus and the next ones that will inevitably arise will stem from strengthened global cooperation frameworks supported by integrated public health AI systems in place.

While sorely lacking that shared framework, there are fledgling initiatives that should become much stronger, more widely adopted and more strategic. The White House put together the most extensive collection of coronavirus literature in one machine-readable, centralized open hub as a call to action for the AI community to help uncover unique insights in the body of data. The EU is fast-tracking the funding of innovation projects that are relevant to dealing with the epidemic. China, which seems to be at a much later stage of the epidemic curve, is sending expert teams and protective equipment to heavy-hit countries. Often-criticized corporations are making immense pledges and rerouting their production for the common good. Thousands of tech startups made their services free to help us cope with a life virtualized in an instant. Mass national-level hackathons to build anti-crisis tech solutions are becoming surprisingly ordinary. Humanity started reorganizing itself with pop-up communities cropping up suddenly, offering solidarity, support, analysis and a chance to jointly rethink the path ahead. Strategic thinking about AI and its adoption will most likely be very much accelerated by the crisis and AI will be instrumental in the post-crisis rebuilding. Globalization in its present form will be rethought and many new models will emerge, balancing global and local.

Isolation can also be splendid – many generations before us were aware of that. If we are granted time to think and to revisit our values, our better nature as a local and global community emerges. At this point, it is hard to estimate which scenario will prevail at the macro level: that of AI technology and human solidarity together defeating the epidemic – or that of a long-lasting nightmare, devastating the economy for a long time and robbing more lives than the world wars did. We remain confident that once the crisis is over, life may not just return to "normal", but that we spend the time and effort experimenting with new models that may resemble the ones laid out in this book, to reshape our modus vivendi as a civilization.

This is an opportunity to reboot, reframe, and redesign our future. And for that, the ultima ratio should be augmented intelligence and humane wisdom, walking hand in hand.

FURTHER READING

Chapter One

Research Priorities for Robust and Beneficial Artificial Intelligence, an open letter; {Stuart Russell, Daniel Dewey, Max Tegmark}, accessible at https://futureoflife.org/ai-open-letter

Chapter Two

Professor Stephen Hawking's speech at the opening of the Leverhulme Centre for the Future of Intelligence accessible at https://www.cam.ac.uk/research/news/the-best-or-worst-thing-to-happen-to-humanity-stephen-hawking-launches-centre-for-the-future-of

Chapter Four

European Commission, 2020 White Paper On Artificial Intelligence - A European approach to excellence and trust available at https://ec.europa.eu/info/sites/info/files/commission-white-paper-artificial-intelligence-feb2020_en.pdf

OECD AI Policy Platform https://oecd.ai/

DARPA, Machine Common Sense program https://www.darpa.mil/program/machine-common-sense

Chapter Five

European Commission Ethics Guidelines for Trustworthy AI

https://ec.europa.eu/futurium/en/ai-alliance-consultation/guidelines/

OECD Principles on Artificial Intelligence

http://oecd.org/going-digital/ai/principles/

World Economic Forum/IPSOS poll on artificial intelligence https://www.ipsos.com/en/widespread-concern-about-artificial-intelligence

Chapter Six

Elon University/Pew Internet Research Center survey on tech and democracy https://www.pewresearch.org/internet/2020/02/21/many-tech-experts-say-digital-disruption-will-hurt-democracy/

Ipsos "Trust: The Truth?" Study September 2019 https://www.ipsos.com/sites/default/files/ct/news/documents/2019-09/ipsos-thinks-trust-the-truth.pdf

Ipsos MORI Millennial Myths and Realities

https://www.ipsos.com/sites/default/files/2017-05/ipsos-mori-millennial-myths-realities-full-report.pdf

Ipsos MORI 2018 Veracity Index

https://www.ipsos.com/sites/default/files/ct/news/documents/2018-11/veracity_index_2018_v1_161118_public.pdf

Ipsos June 2019 Global Advisor survey: Trust in Media

https://www.ipsos.com/sites/default/files/ct/news/documents/2019-06/global-advisor-trust-in-media-2019.pdf

Gilbert, David, November 2019, Vice News 2020 Voters Are Already Being Inundated by Fake News on Facebook https://www.vice.com/en_us/article/bjwky3/2020-voters-are-already-being-inundated-by-fake-news-on-facebook

CIGI-Ipsos Global Survey on Internet Security and Trust

https://www.cigionline.org/internet-survey-2019

Ipsos Global Advisor August 2019

https://www.ipsos.com/en-eg/social-media-should-governments-shut-it-down-times-crisis

Meaker, Morgan November 2019 The Correspondent, How governments use the internet to crush online dissent

https://thecorrespondent.com/142/how-governments-use-the-internet-to-crush-online-dissent/10695281102-64d05e26?fbclid=IwAR0IpCG3r94PxsWvG7--OWCBIJF9s2R-plrRUJW40OVAAfg6FF59Ca40fYg

Ghaffary, Sharin, August 2019, Vox The algorithms that detect hate speech online are biased against black people

https://www.vox.com/recode/2019/8/15/20806384/social-media-hate-speech-bias-black-african-american-facebook-twitter

Roose, Kevin, August 2019, The New York Times, The Making of a YouTube Radical https://www.nytimes.com/interactive/2019/06/08/technology/youtube-radical.html

Stevenson, Alexandra, November 2018, The New York Times Facebook Admits It Was Used to Incite Violence in Myanmar

https://www.nytimes.com/2018/11/06/technology/myanmar-facebook.html

EU Code of Practice on Disinformation https://ec.europa.eu/digital-single-market/en/news/code-practice-disinformation

https://ec.europa.eu/digital-single-market/en/news/last-intermediate-results-eu-code-practice-against-disinformation

BETWEENBRAINS

TAKING BACK OUR FUTURE IN THE AI AGE

AI is a present reality: we live at the threshold of an AI-dominated era. AI is more than a Technology wave. In the 21st century and beyond it is the very core source of power that fuels politics, business, and society: our minds, our work, and our homes. AI also rewrites the rules and forces us to lose no time in rethinking fundamental questions of our humanity. Our human brains will soon be required to adapt at hyper speed to a new paradigm of omnipresent machine intelligence. We all have to deal with both its opportunities and threats in a conscious manner. Presently AI obscurity, hypotheticals, hype, and hysteria are aggravating problems of increasingly polarized and disconnected societies. However, whose image AI development will take and how the AI Age will be shaped is still in the hands of informed and clear-sighted citizens and leaders. The authors undertook the task of making sense of AI, especially its impact on the present and the near future for responsible readers worldwide.

The authors would be delighted to read your messages, shares, reviews and comments. Please reach out and stay in touch for more thoughts:

Web: . . . .
https://www.betweenbrains.ai/

Facebook: . . . .
https://www.facebook.com/betweenbrainsai/

Twitter: . . . .
https://twitter.com/BetweenBrainsAI

Linkedin: . . . .
https://www.linkedin.com/company/betweenbrains/

NOTES

(Endnotes)

1 An Open Letter: Research priorities for robust and beneficial artificial intelligence – Future of Life Institute

2 Stephen Hawking - will AI kill or save humankind?

3 Wherever the first-person singular is being used in the forthcoming chapters, that refers to co-author George Tilesch.

4 Our gratitude goes to Mongolian President Tsakhiagiin Elbegdorj, who kindly shared this lens with us.

5 Google 'ditches contract with us military' after employee revolt, The Independent, June 2018

6 Co-author George Tilesch.

7 Co-author George Tilesch.

8 A significant part of this sub-chapter overlaps with the recommendations in an upcoming Call To Action report by the Club De Madrid AI Policy Working Group: "Trust and Public Debate in the AI Age", led by George Tilesch.

9 Ipsos "Trust: The Truth?" Study September 2019

10 Ipsos MORI Millennial Myths and Realities

11 (Journalists 26%, Ministers 22%, Politicians Generally 19%) Ipsos MORI 2018 Veracity Index

12 Ipsos June 2019 Global Advisor survey: Trust in Media

13 2020 Voters Are Already Being Inundated by Fake News on Facebook, Vice, November 2019

14 CIGI-Ipsos Global Survey on Internet Security and Trust

15 Ipsos Global Advisor August 2019

16 How governments use the internet to crush online dissent, The Correspondent, November 2019

17 The algorithms that detect hate speech online are biased against black people, Vox, August 2019

18 The Making of a Youtube Radical, NYT, August 2019

19 World Economic Forum/Ipsos Global AI Perceptions study July 2019

20 CIGI Ipsos Study 2019 on Internet Security & Trust

21 Facebook Admits It Was Used to Incite Violence in Myanmar, NYT, November 2018

22 IE CGC European Tech Insight 2019

23 Last intermediate results of the EU Code of Practice against disinformation, European Commission, June 2019

24 Demystifying the Black Box That Is AI, Scientific American, August 2017

25 Incivility pushes U.S. toward 'edge of civil war', a new poll says

26 China seeks to root out fake news and deepfakes with new online content rules, Reuters, November 2019

27 New AI fake text generator may be too dangerous to release,

say creators, The Guardian, February 2019

28 Creating a data set and a challenge for deepfakes, Facebook AI, September 2019

29 The opinions and viewpoints expressed in this chapter are solely those of one author, George Tilesch.

30 China is selling autonomous killer drones to the Middle East, Futurism, November 2019.

31 The opinions and viewpoints expressed in this chapter are solely those of one author, George Tilesch.

Made in the USA
Monee, IL
04 November 2026

67383015R00128